GROUP CONFLICT AND POLITICAL MOBILIZATION IN BAHRAIN AND THE ARAB GULF

I0128506

INDIANA SERIES IN MIDDLE EAST STUDIES

Mark Tessler, general editor

GROUP CONFLICT AND POLITICAL MOBILIZATION IN BAHRAIN AND THE ARAB GULF

Rethinking the Rentier State

Justin Gengler

Indiana University Press

Bloomington and Indianapolis

This book is a publication of

Indiana University Press
Office of Scholarly Publishing
Herman B Wells Library 350
1320 East 10th Street
Bloomington, Indiana 47405 USA

iupress.indiana.edu

© 2015 by Justin J. Gengler

All rights reserved

No part of this book may be reproduced or utilized in any form or
by any means, electronic or mechanical, including photocopying
and recording, or by any information storage and retrieval system,
without permission in writing from the publisher. The Association
of American University Presses' Resolution on Permissions
constitutes the only exception to this prohibition.

♾ The paper used in this publication meets the minimum
requirements of the American National Standard for Information
Sciences—Permanence of Paper for Printed Library Materials, ANSI
Z39.48-1992.

Manufactured in the United States of America

Cataloging information is available from the Library of Congress

ISBN 978-0-253-01674-4 (cloth)
ISBN 978-0-253-01680-5 (paperback)
ISBN 978-0-253-01686-7 (ebook)

1 2 3 4 5 20 19 18 17 16 15

To our carefree days in Arabia Felix

"The Battle of Karbala still rages between the two sides in the present and in the future. It is being held within the soul, at home, and in all areas of life and society. People will remain divided and they are either in the Hussain camp or in the Yazid camp. So, choose your camp."

—Ashura banner in Manama, 2006. Quote attributed to Sh. 'Isa Qasim.

Contents

Acknowledgments

Even before my untimely departure from Yemen following the cancellation of the Fulbright program there for security reasons, Mark Tessler had suggested Bahrain as an auspicious candidate for the sort of mass attitude study I hoped to conduct on the topic of group conflict. And when it became clear after some eighteen months of waiting and setback that the project would not be so easily done after all, he continued to offer encouragement and practical advice—to say nothing of his prompt submission of many a fellowship recommendation—that helped ultimately to see the thing through.

The newly-retired Michael Schechter has been for nearly a decade a constant mentor and friend. More recently, since I began working on the manuscript for this book, he has served the helpful purpose of motivator, with his periodic messages asking how many chapters I have still to write and revise. His advice about the selection of appropriate fonts for this volume, on the other hand, was not solicited. David DiPasquale sensitized me to the need for recognizing the real-world policy implications of medieval Islamic political philosophy, and to the usefulness of the new social media as vehicles for information exchange and informed public debate.

My beautiful wife Julia, whom I met in Yemen and later joined in Bahrain, was there to share the ups and downs of my Bahrain field research. Her companionship and persistent optimism helped lighten a process that was otherwise not rarely frustrating. She also provided the final bit of inspiration needed to put an end to the fieldwork once and for all, with the timely delivery of our first child, Maryam, joined more recently by David. Similar thanks are in order for my family who, despite not quite understanding my desire to spend years in Yemen, Bahrain, and elsewhere in a far-flung and seemingly volatile part of the world, were nonetheless steady in their support, and in their much-appreciated willingness to temporarily adopt a cat and a houseful of orchids.

My field research in Bahrain was made possible only through the help of many dedicated, invaluable individuals most of whom, unfortunately, cannot be named here. First among these are my Bahraini interviewers, who braved the heat of the summer, the suspicions and repeated rejections of their fellow citizens, and all for a bizarre survey being conducted by some guy from a university in Michigan. I hope they will take pride in the results of their considerable efforts presented herein. Special thanks go to N. Y. and H., who bore more than their fair shares of this labor and, in the latter case, gave appreciated advice and assistance that went well beyond his role as interviewer. Also indispensable was the regular support of the Public Affairs Office at U.S. Embassy Manama, who helped broker meetings with elusive Bahraini politicians. Finally, I thank the (now former) Bahrain Center for Studies and Research for its sponsorship of my Fulbright fellowship and its many trips to the immigration office to revalidate my entry visa.

Yet my field research in Bahrain also could not have occurred without the preliminary aid of many excellent Arabic instructors at the Yemen Language Center in Sanaʿa, or indeed without the country of Yemen more generally, home to the friendliest, funniest, and most welcoming people I know, never shy to strike up an unsolicited conversation with an odd-looking foreigner. I thank in particular the ever-entertaining ʿAbd al-Qawi al-Muqaddasi and ʿAbd al-Karim al-Akwa, whose love for Arabic Monopoly and televised professional women's tennis, respectively, served to improve my language skills far more than did any classroom sessions.

Numerous organizations provided the financial support to enable all of the above, contributions for which I express my sincere thanks. A Critical Language Fellowship from the U.S. State Department first brought me to the Middle East and showed me how bad my second-year college Arabic really was. For most of the next two years I was able to improve it through a Foreign Language and Area Studies Fellowship from the U.S. Department of Education, an IIE Fulbright Critical Language Enhancement Award, and a David L. Boren Graduate Fellowship. Additional funding for research in Yemen came from an Individual Fellowship from the University of Michigan's International Institute, and a Graduate Student Research Grant from Rackham Graduate School. Support for my field research in Bahrain came from an IIE Fulbright Fellowship; a second Rackham Graduate Student Research Grant; and a Thesis Grant from the Department of Political Science. A Boren Graduate Fellowship enabled me to continue my Arabic instruction while in Bahrain, and the Rackham Graduate School afforded generous financial assistance upon my return from the field. Finally, I am thankful for the continued support of the Social and Economic Survey Research Institute at Qatar University, which has enabled and encouraged the completion of this book, as well as my helpful and perceptive editors at the Indiana University Press, Rebecca Tolan and Sarah Jacobi.

GROUP CONFLICT AND POLITICAL MOBILIZATION IN BAHRAIN AND THE ARAB GULF

Introduction
Bahrain, the First Post-Oil State

THE PERSIAN GULF kingdom of Bahrain is commonly cited as the Arab world's first "post-oil" economy, both in the sense of its being the place of the first discovery of commercial quantities of oil in the region, and also the first to have effectively run out. The former meaning is now largely a point of trivia, the 1932 find by Standard Oil of California (now Chevron) long overshadowed by the far more massive oil and gas reserves subsequently located and exploited in nearby Saudi Arabia, Kuwait, the United Arab Emirates (UAE), and most recently Qatar. Yet, implicit in this designation also is something more than historical fact: the idea that, as first discoverer, Bahrain also was the first representative of a certain *class* of nation, whose members would in the ensuing decades assume a previously unimaginable global significance. This new political genus was, of course, the oil—or rentier—state, kept afloat not through a productive workforce and sound economic management, but by the grace of God (or, less glamorously, by the chance geological distribution of dead plants and animals). Commercial oil had existed for some seventy years prior to its discovery at Bahrain's Jabal al-Dukhan, but the building around it of an entire polity was an experiment never before witnessed.

The other half of Bahrain's designation as the first "post-oil" economy therefore connotes similarly both a factual statement and a cautionary lesson in societal organization and sustainability. Production from Bahrain's 'Awali oil field peaked in 1970 at 76,640 barrels per day, and at the time of its nationalization in 1980 was already in sharp decline. Despite a temporary offset from higher oil prices in the 1970s and early 1980s, it was clear that the site of Bahrain's first oil well would soon live up to its Arabic name: "Mountain of Smoke." When civil war struck Lebanon in 1975, Bahrain seized the chance to diversify away from resource reliance, Manama rapidly replacing Beirut as the financial hub of the Middle East. By 2003, oil output was down 51 percent to a mere 37,550 barrels per day,[1] while a jump in immigration and naturalization to sustain the petroleum and banking industries had augmented the island's population threefold over the same period, from around 215,000 residents in 1971 to more than 650,000 in 2001. In less than ten years that number would again double, with Bahrain's demographic balance tipping ever more to the side of non-nationals. The census of 2010 would find Bahraini citizens for the first time outnumbered, constituting only 46 percent of residents, compared to 83 percent but four decades earlier.[2]

No wonder, then, that Bahrain, with its meager oil revenues and an exploding population, much of which had come to expect the generous welfare benefits of a resource-funded economy—no wonder that this declining rentier state would go on to face greater social and political turmoil than its wealthier Arab Gulf neighbors. The entire system had quite literally run out of fuel. Such is the reasoning, at least, that most often underlies Bahrain's distinction as the first post-rentier state, a political appellation as much as an economic one. And its use is not limited to academic circles. When Arab Spring protests erupted in Bahrain and to a limited extent Oman in February 2011, fellow Gulf Cooperation Council (GCC) states pledged $10 billion in aid to each as part of a so-called "GCC Marshall Plan."[3] It was no coincidence, the suggestion was, that the unrest had hit the council's two smallest oil producers, whose fiscal constraints had permitted the growth of popular resentment.[4] What was needed, ostensibly, was a financial shot in the arm, to shore up the fundamentally economic basis of political activism among Gulf citizens, and so the fundamentally economic basis of stability of the Gulf monarchies.

Bahrain's dubious label is, however, something of a misnomer. While it is true that its domestic oil production has declined precipitously since 1970, in fact revenues from the onshore field have, since a 1958 treaty with Saudi Arabia, represented only a small portion of total oil receipts. In the Bahrain–Saudi Arabia Boundary Agreement, Bahrain ceded its claim to the much larger Abu Saʻafa oil field, located along the maritime border of the two countries, in return for one-half of net proceeds.[5] For almost a decade beginning in 1996, when oil prices breached historical lows, Saudi Arabia even granted Bahrain the full production of the field.[6] By 2003, Bahrain's share of the output from Abu Saʻafa—around 150,000 barrels per day—comprised more than 80 percent of its total production.[7] Ongoing investments to revitalize the 'Awali field have shaved this proportion to around 70 percent as of 2013, with promises of further reductions. Still, absent a radical change in Saudi Arabia's allocation, or in Bahrain's overwhelming dependence upon petroleum exports, Abu Saʻafa oil will account for somewhere between one-half and two-thirds of its total government revenues for the foreseeable future.

One can question, then, whether Bahrain's limited output relative to other oil-producing countries precludes its being a fully-fledged rentier economy. But its status as a post-oil state cannot stem from its having run out of the black stuff, for Abu Saʻafa has rendered its aggregate production quite stable. Moreover, when one puts into context the resource rents Bahrain does receive, it is not clear why the country should be in any worse financial-cum-political situation than some of its neighbors. Certainly, alongside the $351 billion Saudi Arabia earned from oil in 2013, Bahrain's $16.5 billion appears a trifle. But the Saudi take must support a citizen population of over 21 million, compared with barely half a million Bahraini nationals. Saudi Arabia's resource revenues, while vast in raw terms, amount to only around $16,400 per citizen, Oman's only slightly more at around $18,300, and Bahrain's around $29,000. In Kuwait, the United Arab Emirates, and Qatar, by contrast, this ratio is $73,000, $131,500, and a whopping $428,000 per individual, respectively.[8] Thus, if Bahrain's political difficulties result from a state financially incapable of meeting the material expectations of citizens, then pity Saudi Arabia, which ought probably to rethink its role as donor to the GCC Marshall Plan.[9]

To be sure, insofar as Bahrain defies description as a rentier state, it does so not on account of its lack of resources, but for its longstanding inability to transform those resources into the types of social and political outcomes associated with this category. Just as the prospect of indefinite oil wealth was short-lived, so too was that of societal harmony born of a rising economic tide. New employment opportunities in the government sector were a far cry from traditional occupations such as fishing, date palm cultivation, and pearl diving, yet Bahrainis were not on that account content to remain passive subjects, suckling contentedly at the teat of the state. It is a common local observation that Bahrain can go only ten years without a popular uprising, and the near century that has elapsed since the discovery of oil has offered few counterexamples. Absolute tribal power was broken in the early 1920s by British-imposed administrative reforms; since that time, protest and organized opposition have found fuel in Arab nationalism, radical socialism, Islamic fundamentalism, liberal constitutionalism, sectarian rivalry, and labor movements rooted in the very state-owned oil sector assumed to dampen popular political involvement. As Qubain tells as early as 1955, "latent tensions exist that can be exploited if the proper circumstances arise":

> For instance, riots broke out between the Sunnis and the Shi'is during [the Shi'i religious festival of] Muharram in September 1953. It took the use of the whole police force and the imposition of a curfew to bring peace back to the country. In July of the following year, when some Shi'is were convicted for being involved in a fight with Sunnis, the whole Shi'i sect staged a demonstration, conducted an attack on a police post during which four Shi'is were killed, and finally all Shi'i workers went on strike. Six months later, in December 1954, another general strike took place in support of demands made to the government for certain reforms which had already been promised by the ruler.[10]

In affording new ideological and physical grounds for discussion and coordination, then, the petroleum revolution arguably contributed to the rise of public activism in Bahrain. It did not, in any case, effect the opposite.

The February 14th Uprising

In February 2011, encouraged by successive mass uprisings in Tunisia and Egypt, hundreds of thousands of Bahraini citizens took to the streets to call for the ouster of the ruling Al Khalifa family. The date chosen for the start of protests, February 14, marked the nine-year anniversary of Bahrain's 2002 Constitution, a revised charter promulgated unilaterally by the then newly crowned King Hamad bin 'Isa. The document has come to symbolize for regime opponents, in particular for the country's long-disenfranchised Shi'a Muslim majority, the false promise of political reform in Bahrain. Exactly one month after the onset of demonstrations, which saw the violent deaths of both protestors and riot police and prompted a countermobilization by pro-government Sunnis, the movement was finally crushed with the intervention of several thousand ground troops dispatched by neighboring Arab Gulf states eager to contain the mounting crisis.[11]

This book is not the story of that uprising—not, at least, in the immediate sense. It was in the making long before protesters occupied the now-flattened Pearl Roundabout and renamed it "Martyrs' Square." Of course, in describing the conditions that gave rise to Bahrain's failed revolution, it does offer a framework through which to view this latest episode in a tumultuous political history. Yet its real purpose lies elsewhere, and the net it aims to cast is far wider. Though its primary focus is Bahrain, the investigation here seeks to examine a larger class of cases of which this tiny archipelago in the shallow waters off Saudi Arabia is but the best contemporary example. This category one might call the failed rentier state: a state flush with historical levels of resource revenues, yet unable to buy the political acquiescence of its citizens—or, of a particular sort of citizen. That such a government is unable to do so is a problem not only for itself, but also one for political science, whose standing interpretation of the Arab Gulf monarchies revolves precisely around this presumed ability to appease would-be opponents through material benefaction. Should there exist, then, identifiable circumstances under which this formula for political buy-off does not obtain, we must revise not only our expectations about the inherent political stability of the Arab Gulf regimes, but also our understanding more generally of the nature of political life in rent-based societies.

This work elaborates one important qualification to the premise that economically satisfied Gulf Arabs make politically satisfied Gulf Arabs: the existence of societal division along ascriptive group lines, whether ethnic, regional, tribal, or, as in the case of Bahrain, denominational. Utilizing insights from Bahraini political leaders and the results of an original, nationally representative survey of mass political attitudes, it demonstrates how ascriptive identities offer a viable basis for mass coordination in a type of state thought by its very nature to lack one. The empirical analysis shows that the political views and behavior of ordinary Bahrainis are not determined primarily by material considerations, but by one's con-fessionally defined position as a member of the political in- or out-group. What is more, it reveals how the material benefits conferred by rentier states are not dis-tributed in a politically agnostic manner, but aim primarily to reward supporters rather than convert opponents. Hence, in Bahrain and other Gulf societies in which ascriptive categories are politically salient, neither is the rentier state willing to offer its presumed material wealth-for-silence bargain to all citizens, nor are all citizens willing to accept it.

Within this critique also is a larger lesson: that the nature and strength of the individual-level link between economic satisfaction and political quiescence in rent-based economies will depend necessarily on the strategy of rulership adopted by the regime in question. If the relationship between citizen and state is based entirely on economic patronage, and if such patronage is extended to all citizens universally, then the political result may indeed be something akin to the classic rentier state, which need worry only about preserving a minimum level of material satisfaction among subjects. If, however, economic distribution is only a part of a state's wider strategy of political legitimization, or is largely confined to a certain subset of the population, then one should have a different set of theoret-ical expectations. In short, the differing strategies of rule witnessed today across the Gulf monarchies must translate into equally divergent expectations about the extent to which, and among whom exactly, rentier mechanisms hold true.

Testing the Untested

A curious fact about the proposition that economic satisfaction breeds political indifference in resource-dependent states—about this "rentier state theory"—is that for a conceptual framework first proposed some three decades ago and pop-ular ever since, it has yet to be put to the test empirically. Certainly, some of its corollaries have invited quantitative research, most notably its implication that, at the country level, the extent of a nation's resource dependence should tend to be inversely related to its degree of democracy, since more income at the regime's disposal means more citizens content to relinquish their political prerogative in exchange for material reward. Other studies proceed one step further to associate democracy with macroeconomic proxies for rentierism, such as rates of taxation and government-sector employment.

Yet, for all their effort, these analyses cannot bring us closer to demonstrat-ing the individual-level link between material contentment and political apathy—the theoretical glue holding together the rentier framework—precisely because such analyses do not operate at the individual level. That the regimes of the Arab Gulf are both autocratic and resource-dependent does nothing to show that, in 2015 in the United Arab Emirates, or in Kuwait, individual citizens who are satisfied with their economic situation also tend to be satisfied with their country's political situation—or at least uninterested in changing it. Equally, that Saudi Arabia and Qatar maintain high public employment rates and do not impose income taxes cannot directly connect the individual-level economic outcomes of these policies to citizens' political orientations. In short, extant empirical evaluation of the rentier hypothesis has been limited overwhelmingly to tests of the very observations that gave rise to the theory in the first place, while its own proposed causal logic goes unexamined.

For, at its core, the rentier state thesis is less a story about the political machinations of greedy governments than it is about human nature and its impact on individual political behavior under certain conditions. Indeed, the most provocative claim of rentier theory is exactly this, that it purports to understand the very political motivations of citizens: why it is that people become involved in, or alternatively shrink from, politics; what it is that leads one to support, quietly accept, or actively reject a largely unaccountable government. Economics, it suggests, is king; competing factors, it says by omission, must take a back seat. From here is it plain that any proper assessment of the rentier state framework must investigate what the latter professes already to know: the individual-level determinants of political views and behavior in highly clientelistic, rent-based societies. And as it was the Gulf region that served as archetype for the rentier paradigm, it is perhaps only fitting that its first real test should be conducted here.

But such a thing is easier said than done. Macro-level indicators measuring resource exports, political openness, and rates of taxation and public-sector employment are readily available for most countries of the world; reliable data on the political opinions and behavior of ordinary Gulf Arabs are emphatically not. In addition to a political environment traditionally hostile to public opinion research, and particularly hostile to research that would elucidate popular political opinions and societal demographics, the dearth of survey data from the Gulf stems also from more practical causes. Sampling frames are either wanting entirely or treated as secret by state statistical authorities. In any case, few local institutions enjoy the capacity and freedom to undertake scientific data collection. Meanwhile, target populations are at best disinterested in, and more often wary or suspicious of, survey research. As a result, even the two foremost initiatives to compile cross-national data on mass political attitudes globally and in the Arab world—respectively, the World Values Survey (WVS), begun in 1990, and the Arab Democracy Barometer (AB), launched in 2005—have succeeded despite their considerable efforts and resources in surveying the Arab Gulf states but four times as of the time of writing. And none of these surveys managed to field the crucial but highly sensitive questions about normative political opinions and political activities.

Yet, even if one were to obtain such individual-level data, what is it exactly that one would expect to find? Why should one doubt the abilities of the Gulf monarchies to purchase political stability by distributing resource royalties to citizens? With the exception of Bahrain, the Arab Gulf as a distinct category of nations seems to have succeeded in avoiding the sort of mass discontent that toppled or continues to threaten regimes across the Middle East and North Africa. And, not coincidentally, all Gulf rulers have appealed to citizens' wallets through generous social welfare packages announced soon after the Arab Spring arrived in the region. Again, therefore, what gives one reason to believe that the rentier state interpretation does not more or less accurately capture Arab Gulf politics?

The answer, I argue, turns around one's interpretation of Bahrain. If one views the country's defiance of basic rentier assumptions—of citizen disinterest in politics, of a lack of organized political opposition, and ultimately of regime stability—if one believes such contradictions the result of a Bahraini domestic politics that is *sui generis* among the Gulf states, then, certainly, its lessons for the region and for political science are limited. Either Bahrain's rulers are singularly inept at political co-optation, they alone lack the resources to accomplish it, or Bahrainis are uniquely recalcitrant among Gulf peoples. But if, on the other hand, the conditions that underlie Bahrain's dysfunction apply in degrees to the other societies of the region; if its perennial political crises represent not a theoretical exception but merely the realization of a latent possibility that exists in other Arab Gulf regimes according to their peculiar vulnerability to such conditions, then the example of Bahrain is far more instructive. Insofar as there exist identifiable circumstances under which the standard rentier interpretation of Gulf politics is not valid, circumstances that describe Bahrain particularly but not uniquely, then through studying this case one may not only arrive at a necessary theoretical revision, but also a better practical understanding of citizen-regime—and, as will become important, citizen-citizen—relations in the Arab Gulf region and beyond.

Rethinking the Rentier Bargain

The contemporary record of Gulf politics shows that Bahrain is not alone in witnessing a seeming breakdown of the wealth-for-acquiescence agreement supposed to operate in rentier societies. In fact, one need not even reference the empirical failures of the rentier state model to understand why such an open-ended bargain between rulers and subjects never existed at all. In the first place, as opposition activists across the region today can attest, not all citizens will be persuaded to forfeit their political prerogative by the promise of material wealth—or, for that matter, by the promise of physical violence. Certainly, one can imagine myriad sources of political motivation independent of economic concerns: perceptions of group-based discrimination and inequality; a desire for representative and democratic governance as an inherent good; or adherence to revolutionary ideologies such as Arab nationalism, socialism, or, more commonly today, transnational Shi'ism, Salafism, and pan-Islamist currents such as the Muslim Brotherhood.

Moreover, and more fundamentally, even if a state could buy the unanimous support of citizens, it need not even attempt to do so, for it requires only a minimum coalition of supporters with the physical (military) preponderance sufficient to protect it from potential challengers. Indeed, why waste limited resources chasing citizens opposed to the status quo when they might be used to reward those who already have a material stake in its preservation? Rather than deploy limited resources inefficiently upon the whole of society, rulers of distributive states such as those of the Gulf generally seek to maximize their own share by rewarding disproportionately a finite category of citizens whose support is sufficient to keep them in power, while the remaining population is comparatively excluded from the private rentier benefits of citizenship. This incentive for targeted distribution is especially great in countries where a sizable national population (e.g., Saudi Arabia) and/or low resource revenues per citizen (Bahrain, Saudi Arabia, Oman) would limit the political utility of a more egalitarian allocation. Yet, even states less constrained by such factors (Qatar, Kuwait, the United Arab Emirates) have sought to segment their political markets, erecting tiers of citizenship that confer discrete levels of benefits and that imply, in turn, varying levels of economic-cum-political clientelism.

Finally, beyond the diverse nonmaterial concerns of citizens and state incentives for unequal distribution, yet another process working to undermine the individual-level link between economic and political satisfaction in the Arab Gulf states is the ongoing effort by GCC governments to diversify their sources of political legitimacy. Whereas much analytic focus continues to be directed at the problem of *economic* diversification away from reliance upon resource rents and sprawling public sectors, far less examined has been the parallel effort by Gulf rulers—one they have arguably approached with much more seriousness than the former question—to undertake *political* diversification away from purely economic bases of legitimacy. Such a strategy has assumed various forms across the region, but commonalities include a focus on "national" culture and heritage nurtured and protected by the person of the ruler; opportunities for higher education and personal self-fulfillment supplied often by Western institutions; appeals to religious or tribal legitimacy; the pursuit of international prominence and prestige; and the provision of (and the highlighting of the provision of) other intangible benefits such as political *stability* over against political *accountability*.

This latter argument has acquired particular force in the wake of the Arab uprisings begun in 2010—not, paradoxically, because the Gulf states largely escaped the upheaval witnessed elsewhere, but precisely because many, indeed most, did not. Ruling families in Bahrain, Saudi Arabia, Kuwait, and most recently the United Arab Emirates have pointed to the chaos abroad to explain and justify the need for measured reform at home and to mobilize popular support against those who would dare to upset the comfortable if perhaps non-democratic political status quo. The Shiʻa of Saudi Arabia and Bahrain, the mainly Sunni tribal opposition in Kuwait, and members of the Muslim Brotherhood in the UAE—such groups represent at once political scourge and political boon: the potential or actual basis of organized opposition, but also a bogeyman with which to rally the rest of society—or, at the least, frighten it into inaction. Especially in the post-Arab Spring period, then, those Gulf states in which political boundaries follow religious or other ascriptive group lines have hit on a powerful if precarious new source of noneconomic legitimacy: not simply the provision of stability in a region gripped by chaos, but the veritable protection of citizens—*loyal* citizens—from feared enemies abroad and their subversive domestic agents. Simmering tension and outright hostilities between the Sunni Arab monarchies and Shiʻa-led regimes in Iran, Iraq, and Syria have only reinforced this sectarian narrative.[12]

The First Mass Political Survey of Bahrain

This proposition, that group competition helps explain not only Bahrain's inability to buy political assent, but also the observed discrepancy between rentier theory and rentier reality throughout most of the Arab Gulf, presents no lack of challenges for the one looking to demonstrate it empirically. For one requires not only individual-level data on the political attitudes and economic conditions of ordinary Gulf Arabs, but moreover demographic data identifying individuals as a member of one or another ascriptive community. And when even the aggregate proportions of Sunnis and Shiʻis within GCC states is a matter of speculation—as governments refuse to provide, or more often claim not to collect, such statistics—these two requirements render inadequate all extant data sources, including the aforementioned World Values and Arab Barometer surveys.

Hence, in early 2009, I endeavored to collect new data from Bahrain, undertaking the first-ever mass political survey of Bahraini citizens. Administered to a nationally representative sample of 500 random households spread across the island, the survey employed the standard Arab Barometer instrument that investigates a wide range of social, religious, and political behaviors and attitudes, in addition to household economic data. The Bahrain survey also recorded respondents' confessional affiliations, allowing not only an empirical test of the present argument, but also the first window into the country's confessional demographics in almost seventy years. The last time the government of Bahrain reported official statistics on its Sunni and Shiʻi communities was in its very first census of 1941.[13] Utilizing these previously unavailable Bahrain data, this book presents the first systematic empirical assessment of the individual-level assumptions underlying the rentier state framework.

Beyond its impact on politics per se, among the transformative ideational effects of the Arab Spring was to deal a coup de grâce to the ubiquitous notion of the Gulf "oil sheikh," happy to abandon his country to rule by princes and Islamic extremists while he revels in a life of luxury made possible by his personal cut of the nation's oil revenues. Not just in Bahrain but all across the Arab Gulf region— in Saudi Arabia, in Kuwait, and to a lesser extent in Oman and the United Arab Emirates—observers witnessed a surprising cache of political enthusiasm among ordinary people, and this directed against governments that had supposedly co-opted their support with oil. This book offers not only a new conceptual framework to make sense of this seeming contradiction, but also, and for the first time, empirical evidence to back it up.

Summary of Chapters

The book proceeds in six chapters. Chapter 1 offers a more expansive account of the conceptual framework introduced already: a theory of group-based political mobilization in the Arab states of the Gulf. Challenging some four decades of received wisdom from rentier state theory, it argues that the region's unique political and economic institutions do not serve to preclude mass political coordination, but rather to privilege a certain *type* of political cooperation, namely coordination on the basis of outwardly observable social categories such as ethnicity, religion, tribal or regional affiliation, and so on. The result is a structural tendency in the Gulf toward ascriptive group politics, which in countries with diverse and/or regionally diffuse populations tends to induce political contestation not by citizens competing simply for additional allocations of state benefits *qua* individuals, but by larger ethnic-cum-political groupings competing both over material benefits as well as over control of the polity itself.

Moreover, it argues, rather than a universal rentier social contract, whereby the rulers of rent-based economies endeavor to purchase the universal political support of citizens using economic benefits, instead Gulf rulers themselves face strong incentives to segment their political markets along these same descent-based groupings, to *reinforce* rather than de-emphasize society's latent social distinctions. Rather than deploy limited resources inefficiently upon the whole of society, the controllers of rentier states instead seek to maximize their own consumption of the material benefits of rule, by offering citizens only the minimum allocation necessary to ensure a winning coalition of supporters. Thus, the primary task of rentier governments is not the distribution of resource wealth to the population merely, but doing so as cheaply as possible.

As such, the most elementary question concerning the rulers of rentier societies is how to achieve the optimal balance between economic and political autonomy, that is, how to maximize simultaneously both (a) private enjoyment and discretionary employment of directly accruing resource wealth; and (b) freedom from popular accountability through economic appeasement via distribution.

Among the diverse solutions to this rentier dilemma is a strategy of political segmentation, in which a state disproportionately rewards a class of citizen supporters and disproportionately excludes the remainder from the rentier benefits of citizenship. In this way, material benefits are not dissipated across the whole of society but concentrated on a finite constituency whose support is sufficient to ensure the continuity of the regime. The contemporary example of Bahrain represents the purest case of this strategy of purposeful societal division.

Chapters 2 and 3 give additional substance to this theoretical account by studying the case of sectarian political mobilization in Bahrain. Drawing insights from interviews with some dozen Bahraini political and religious leaders—four of whom are now serving lengthy prison terms for their roles in the February 14th uprising—this section describes how, in Bahrain, the individual-oriented politics of economic competition assumed to operate in allocative states is superseded by a group-based contest to determine the very character of the nation itself: its history and cultural identity, the bases of citizenship, and the conditions for inclusion in public service. It details further how, due to this domestic and regional Sunni-Shi'i competition, the unique political pressure-relieving tools said to be available to Bahrain as a rentier economy are rendered inefficient or even inoperable.

Chapter 4 supplies a practical and methodological preface to the analysis of the Bahrain mass survey, detailing the actual survey procedure, addressing theoretical and methodological questions, and offering a first direct, reliable look at Bahrain's confessional demographics since its 1941 census. Chapter 5 employs the Bahrain data to evaluate the book's main theoretical claims. It begins by examining patterns of public and private goods provision in Bahrain, particularly with respect to employment in the state sector. It next assesses the relative roles of economics versus group dynamics in determining Bahrainis' political orientations and behavior. It uses the individual-level survey data to test whether citizens' normative attitudes toward the state, as well as the political actions they take for or against it, are influenced foremost by material satisfaction, as per the rentier state hypothesis, or by sectarian religious affiliation and orientations.

A concluding chapter 6 reviews the preceding, makes note of its limitations, and suggests how the inquiry might be extended as part of a larger revised Arab Gulf research agenda that both reflects more closely the political trends observed in the region and makes better use of new methodological resources presently becoming available.

1 Group-Based Political Mobilization in Bahrain and the Arab Gulf

Born of the newfound importance of oil-exporting nations in the 1970s and 1980s, the idea of the "rentier economy" arose in economics as a description of those countries that rely on substantial external rent, the latter defined broadly as a reward for ownership of natural resources, whether strategically located territory, mineral deposits, or, more to present purposes, oil or natural gas reserves.[1] A special category of the rentier economy, a "rentier state," came to describe those economies in which only a few are engaged in the generation of this rent, the archetypal examples of which were, and remain, the oil- and gas-rich monarchies of the Arab Gulf. In rentier states, then, the creation and control over wealth is limited to a small minority of society, that is to say, to "the state," or, in the case of the Gulf regimes, to the ruling family *qua* state, while the vast majority of residents and citizens play the role either of distributor or consumer.

With such an extreme economic-cum-political imbalance thus written into the very definition of the rentier state, a significant portion of the rentier literature concerns itself with an inherent puzzle: how are these regimes seemingly so durable? That is, why do the citizens or residents of rentier states not simply confiscate the revenue-generating resources from their physical owners? The latter, after all, are hopelessly outnumbered. Yet, far from the gloomy predictions about the post-colonial fates of the Gulf monarchies, more than 40 years after the British withdrawal from its protectorates in 1971, the Arab Gulf states "continue to be ruled by the same families, sometimes the same individuals, under the same traditional forms and within virtually the same borders that had been engineered by the British political agents as they departed."[2] And all this despite having seen three major regional conflicts, several oil crises, the rise and fall of Arab nationalism, the Iranian Revolution, the Arab Spring, and the threat of similar revolutionary episodes across the Gulf. What has enabled—continues to enable—these seemingly anachronistic regimes to survive?

A Hard Bargain: The Classical Origins of the Rentier State

Beginning with the earliest statements of the rentier state framework, theorists have posited that the resource-controlling parties within rentier states can, in short, buy off would-be domestic opponents through judicious economic policy. The form of such policy is at once positive (rent controllers offer citizens a portion of their wealth as public and private goods) and negative (they agree *not* to expropriate from citizens as they otherwise would like to). In practical terms, these avenues of mass co-optation correspond to two complementary mechanisms by which modern Gulf regimes are said to use their economic hegemony to elicit political acquiescence. First, they employ those who need employment; and, second, they abstain from levying taxes "on the basis," in Vandewalle's oft-repeated formulation, "of the reverse principle of no representation without taxation."[3] Together these incentives foster a rent-induced consensus that "helps explain why the government of an oil-rich country . . . can enjoy a degree of stability which is not explicable in terms of its domestic economic or political performance."[4]

"Every citizen" of a rent-based economy, Beblawi insists, "has a legitimate aspiration to be a government employee; in most cases this aspiration is fulfilled."[5] Judging by present public-sector employment rates in the Arab Gulf, it would seem that this idea is as valid today as it was decades ago. In 2009, the National Bank of Kuwait estimated that Gulf citizens working in the public sector "account for 58% of total GCC nationals employed," including 50 percent of Saudis, 84 percent of Kuwaitis, and almost 90 percent of Qataris."[6] Among respondents in my 2009 Bahrain survey, around 43 percent report being employed in the public sector. These already vast proportions, moreover, are on the rise: public sector jobs in the GCC rose at an average of 5.2 percent per year during 2006 and 2007, with Qatar recording a spectacular 33 percent annual growth rate, followed by the UAE and Oman at 5 percent, Kuwait at 4.4 percent, Saudi Arabia at 2.9 percent, and Bahrain at 2.4 percent.[7] More recently, bolstered social welfare spending in response to the Arab uprisings has only pushed these percentages higher. In Oman, for instance, private-sector employment among nationals decreased by 4 percent in 2011 over the previous year. In Kuwait, the number of nationals entering the public sector in 2011 almost doubled, prompting the country's finance minister to warn in March 2012 that public-sector wages now equaled 85 percent of oil royalties.[8]

By establishing an entanglement of bloated government ministries, subsidizing large, state-owned conglomerates, and spending huge sums on disproportionately large and well-equipped militaries, Gulf regimes can sop up a young populace that is easily disaffected, eager to marry and find housing, and generally well educated yet nonetheless ill-equipped for work in the private sector. The upshot is that the latter will be content to live their days as government pensioners and social welfare recipients, careful not to kill the goose that lays the golden egg. For their part, ruling families gain a political ally—at worst a self-interest-maximizing, apolitical animal—and need forfeit only a portion of their rent proceeds to guarantee continued enjoyment of the remainder.

Yet the state's beneficence does not end there. At the same time that it subsidizes citizens via employment, it also agrees not to extract through taxation: the rentier gods both giveth and doth not take away. "The cry of the American Revolution," writes Gause "was 'no taxation without representation.' None of the Gulf governments seems willing to take the political risk that direct taxation entails."[9] As Ayubi aptly summarizes,

> The taxation function is thus reversed in the oil state: instead of the usual situation, where the state taxes the citizen in return for services, here the citizen taxes the state—by acquiring a government payment [i.e., a salary]—in return for staying quiet, for not invoking tribal rivalries and for not challenging the ruling family's position.[10]

Here, then, is the ostensive political bargain that has allowed the unforeseen longevity of rentier states, and the Gulf monarchies in particular, since their rise to prominence over the last half-century: ordinary citizens are content to forfeit a role in decision making in exchange for a tax-free, resource-funded welfare state. By this conception, economic satisfaction is the primary variable influencing the extent of popular political interest and expectations of participation in decision making, with other, nonmaterial factors playing no important systematic role at the individual level.[11]

In recent years attempts have been made to update the rentier state paradigm to make it less deterministic and to account for important changes in the structure of both the Gulf state and Gulf society since the theory's original articulation. Still, such revisions have, with very few exceptions,[12] not questioned the fundamental material basis of the citizen-state relationship.[13] The original conception of a state unconstrained by social pressures may have been qualified and moderated, yet the basic rentier formula continues to compute in the popular and scholarly imagination: economic subsidies in exchange for political subsidies. Citizens are overpaid relative to their actual productivity, and in return governments are overpaid in political support relative to their performance.

Studying the Rentier State

If statement of the rentier state hypothesis is straightforward enough, however, the matter of testing it—of evaluating in an objective and scientific manner the extent to which its predicted causal mechanisms in fact operate in actual rentier states—is a far less simple exercise. For extant empirical evaluations of the rentier framework suffer from basic theoretical and methodological limitations such that one might argue the theory has yet to be tested at all. Rather than evaluate the central rentier hypothesis, the proposition that material satisfaction breeds popular political apathy and thus political stability in a particular class of state, contemporary scholars have used instead the macro-level causal mechanisms identified in the foundational rentier literature—resource revenues, government expenditures, and taxation rates—to explain an altogether different phenomenon.

Beblawi and Luciani propose cogently that the rentier hypothesis "helps explain why the government of an oil-rich country . . . can enjoy a degree of stability which is not explicable in terms of its domestic economic or political performance." But in place of state *stability*—the political outcome of principal interest to the early framers—contemporary theorists have inserted their own modern preoccupation with *democracy*.

Beginning with Ross's landmark 2001 article *Does Oil Hinder Democracy?*, quantitative tests of the link between resource rents and democracy have dominated the rentier literature. What is more, almost all of these works have utilized exactly the same dependent variable: the ubiquitous Polity IV −10 to 10 scale of regime type.[14] The difficulty with this procedure is twofold. First, there is, as one would expect, little within—or between—country variation in this measure among the Arab Gulf states: Saudi Arabia and Qatar are rated −10 for each year of their existence; the UAE is a perennial −8; modern Oman ranges between −10 and −8; and Bahrain and Kuwait from −10 to −7.[15] At the same time, the fuel rents of the six GCC states exceed the rest of the world by two orders of magnitude: according to Ross's own data from a 2008 paper on oil and gender equality,[16] the mean per capita fuel rents among GCC states is $11,339, compared to $270 for the other 163 countries in the sample. For these two reasons, most of the variation in "democracy" attributed to "oil" should, in truth, be attributed to the Gulf only. In which case we find ourselves in the same position in which we began, namely faced with the question of how to understand the unique political economy of a finite category of states.

A simple plot of these two variables for the full sample of 170 countries clearly reveals the methodological issue underlying attempts to associate resource rents with democracy in the customary manner. Depicted in Figure 1.1 is the relationship between a nation's 2007 Polity IV score and Ross's 2008 fuel rents per capita measure.[17] One will notice, first, that only a small proportion of the countries are identifiable owing to the large cluster of observations hovering at the far end of the *x*-axis. Of those that do stand out, moreover, six are the Arab Gulf states, shaded in gray for ease of identification; the two other outlying cases are Brunei ("BRN") and to a lesser extent Libya ("LBY").

One sees therefore how, despite standardization of the rents per capita measure, the extreme between-country variation in rent-generation—that is to say, the vast difference separating rentier and non-rentier economies—obscures the true system-level relationship between resource rents and democracy. Indeed, it is evident that the bivariate least-squares regression line describing this relationship, which purports to show an immensely significant negative association between a country's per capita rents and polity score, is almost entirely dictated by the small number of outlying observations consisting of the Arab Gulf states along with Brunei and Libya.

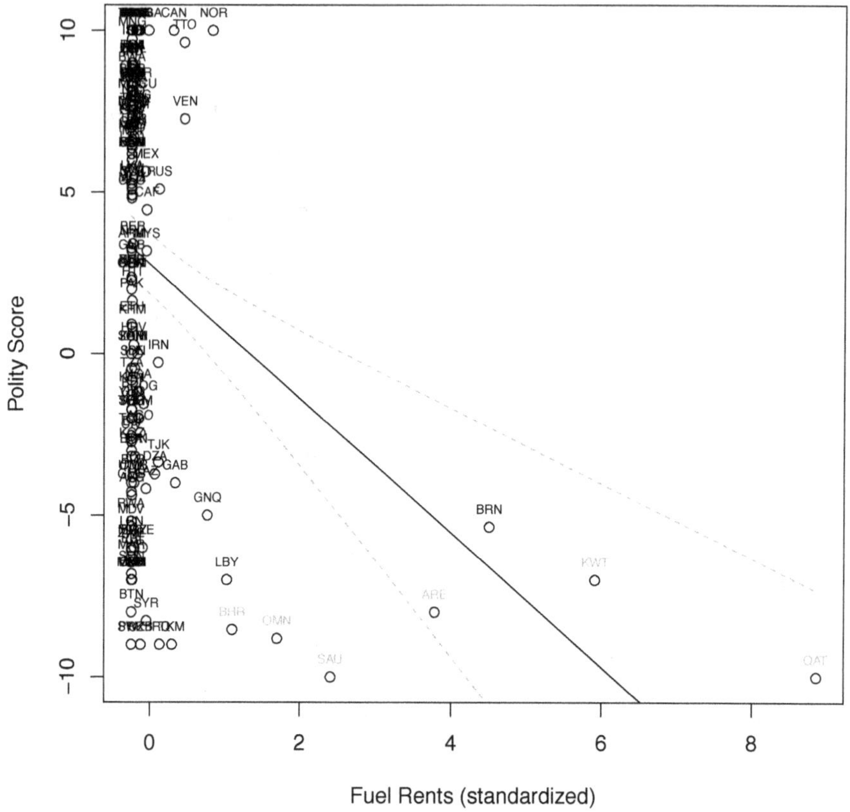

Figure 1.1. Fuel rents per capita and democracy (from Ross 2008).

When one omits these eight outliers one finds that the picture, though more in focus, is still far from clear. Figure 1.2 illustrates the results of this exclusion. Although the regression line describing the estimated relationship between fuel rents and regime type remains apparently negative, its slope is no longer statistically distinguishable from zero. In fact, as indicated by the dotted upper and lower bands of the 95 percent confidence interval, one is unable to rule out the possibility even that the true bivariate relationship is positive rather than negative.[18] It might then be said that the most common application of rentier state theory in political science today, as an explanation for the lack of democracy in resource-rich nations, not only errs in its choice of dependent variable, but also, in doing so, paradoxically draws one back to the original task of rentier theorists: understanding the political ramifications of a mode of economy unique to a finite group of nations. At bottom, Figures 1.1 and 1.2 demonstrate how the category "rentier" exists as a class of state of which one either is or is not a member, as per Luciani's

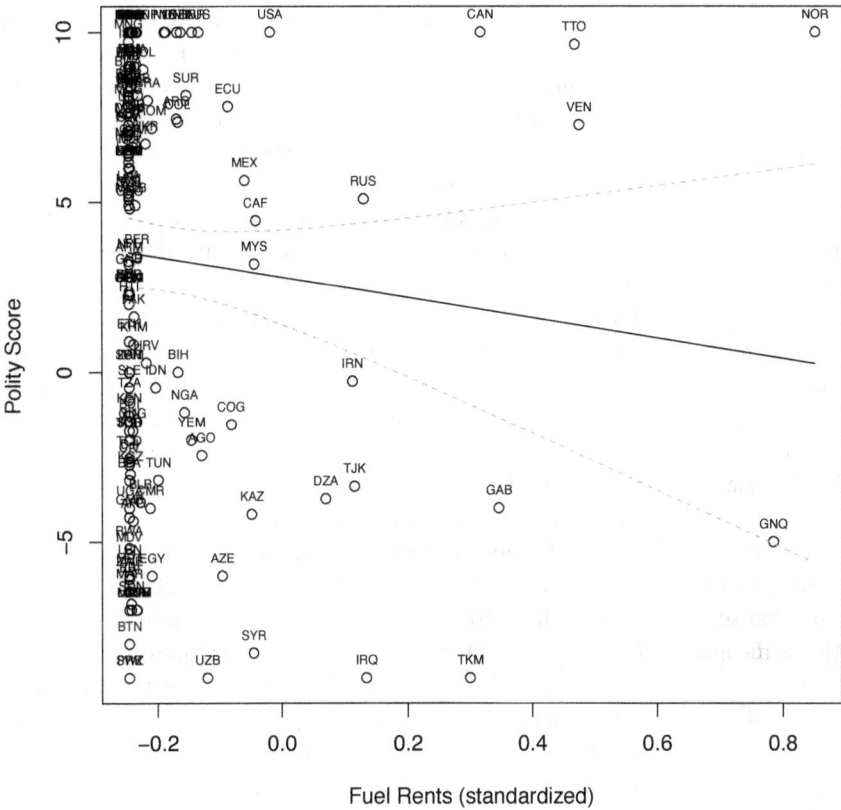

Figure 1.2. Fuel rents per capita and democracy, excluding outliers.

dichotomy of allocative versus productive states.[19] The mystery, accordingly, is not whether an additional dollar of oil profits begets some marginal shift toward authoritarianism in Denmark or New Zealand, but whether it indeed is true that in a handful of rentier societies, six of which are the Arab Gulf monarchies, citizen-state relations operate in a qualitatively different manner than they do elsewhere.

This discussion also suggests the more fundamental, theoretical problem affecting extant attempts to demonstrate the empirical validity of the rentier state framework. Simply stated, they fail to test the actual individual-level causal processes that the theory posits. It is, after all, quite explicit in claiming that the reason states with sizable external rents tend to remain stable (yet authoritarian) is because ordinary citizens, when satisfied economically, are content to concede the realm of politics to their benefactors. Rather than evaluate this specific causal hypothesis, however, investigators have sought to link country-level economic variables to country-level political outcomes such as regime type or democratic transition. Yet such studies can, at best, only confirm the existence of these macro-level associations; absent a new theory that ties the latter together directly without recourse to the individual level of analysis, they bring one no closer to knowing whether the rentier model is correct in its account of what underlies these links. For the theory's boldest statement is not what it says about rent-dependent states themselves, but what it assumes about their citizens: that it understands the drivers of popular interest and participation in politics; what it is that inclines ordinary citizens to seek an active role in political life or, alternatively, to shrink from it. These are no small claims.

Of course, the form of previous empirical testing was determined in part by the nature of available data. And such data, owing to numerous practical hurdles, have not been informed by mass surveys of the political attitudes of ordinary Arabs—to say nothing of Gulf Arabs—until very recently, and even then on a limited and sporadic basis. Thus, the failure of prior studies to test the individual-level causal story underlying the rentier framework is not simply a product of theoretical or methodological oversight. However, with the completion in 2014 of the third wave of the Arab Democracy Barometer survey project, as well as contemporaneous WVS surveys in Qatar and Kuwait, further neglect of this inquiry now that such an opportunity exists would represent continuation down a path that is incapable, ultimately, of answering the most elemental questions to which we seek answers: What causes individuals to incline toward, or abstain from, politics in the rentier states—indeed, in the most emblematic and, in practical political terms, the most important of all the rentier states? Is the prevailing explanation correct in identifying material well-being as the key determinant? If so, is this relationship between personal economy and personal political orientation a universal one, or does it obtain only under certain conditions? in certain countries? or for certain types of individuals?

One might therefore dare to say that the theoretical architecture of the rentier state paradigm first described in the 1970s and repeated until today—the wealth-for-silence bargain extended to citizens of rent-based regimes—in fact has never actually been evaluated empirically. For all the studies that have since purported to do so, insofar as these have examined associations between country-level phenomena rather than analyze the link between material well-being and political involvement among individual citizens, like the science of gravity these have tested only the outwardly observable interactions of the rentier state rather than its internal causal processes. And, to be sure, the difference is not inconsequential. Not all states have been equally successful at converting external rents into domestic stability, there being important cross-country variation among the Arab Gulf states that one cannot explain without a clearer understanding of how interceding variables at both the country level and the individual level serve to condition the relationship between politics and economics in rent-dependent societies.

Revisiting the Rentier Bargain

Now more than ever before, there exists a pressing need for a systematic revisiting of the rentier thesis as a blueprint for understanding Arab Gulf politics. For, quite apart from the question of rigorous *empirical* testing of the existing theory, the contemporary political record would seem to demand a basic *theoretical* reevaluation of the framework itself—a rethinking of the oft-cited "rentier bargain" said to exist between Gulf rulers and their citizens-cum-clients. Not only have ordinary people across the region revealed an unexpected wellspring of political enthusiasm, but they have done so—and, continue to do so—at a time when ruling families enjoy unprecedented levels of resource rents thanks to skyrocketing oil prices, an historic windfall enabling vastly inflated if likely unsustainable state budgets. As Okruhlik observed already in 1999, "[I]n Saudi Arabia, Kuwait, and Bahrain opposition has arisen and with it a discrepancy between the expectations derived from the rentier framework and empirical reality."[20]

But what to make of this popular inclination toward political life in the Arab Gulf, contradicting as it does so much received wisdom? One may raise two basic possibilities, with qualitatively different implications for the efficacy of the rentier state model. First, political activism may be a result merely of structural limitations upsetting the system of patronage linking rulers and ruled. Public sectors may now be too saturated to employ additional would-be supporters, for instance; immigration rates too high to allow ample provision of quality housing, schools, health care, and other services; or security budgets too vast to finance competing public and private goods. In which case the question is one of policy: to diffuse heightened political pressure, states might improve education to produce graduates who qualify for private-sector positions and thus need not look to the state for employment, stem the incessant inflow of cheap migrant labor, or trim military expenditures to less than their current world-leading levels.

A second explanation, however, is that the rentier paradigm itself has misjudged the drivers of political interest among Gulf Arabs, or has not defined them clearly in the first place. If concerns over economic well-being are not the sole or even dominant factor shaping the political orientations and behavior of Gulf citizens, or are dominant in some contexts and under some circumstances but not in others, then the issue at stake is far more fundamental and reopens the inquiry into how politics actually operates across the region. In which case standard refrains such as "no taxation, so no [basis for] representation" may be analytically familiar and convenient descriptions of the citizen-state relationship in the rentier Gulf, but not necessarily accurate ones.[21]

Unfortunately for proponents of a predominantly economic explanation of Gulf politics, even a cursory survey of political opposition and mobilization throughout the region, not least in the aftermath of the Arab uprisings, is enough to show that while economic concerns—corruption, inequality, unemployment—have played a role in generating popular support for reform of the prevailing monarchical regimes of the region, even more critical as motivating and especially *mobilizing* forces have been issues surrounding group identity and conflict: distinctions along the lines of religious denomination, region, tribal versus nontribal, and Islamist versus secular. Even in those instances in which political disagreement has stemmed in part or in whole from socioeconomic grievances, with few exceptions[22] this conflict has manifested itself both rhetorically and institutionally along ascriptive group boundaries.

In Saudi Arabia, a reform movement once limited mainly to the kingdom's Shi'a minority sees increasing appeal among other citizens structurally excluded from the political and economic benefits enjoyed by the Najdi tribal ruling elite. In Kuwait, a longstanding urban-tribal divide has, since the Bahrain uprising, increasingly overlapped with sectarian cleavages to provoke a series of constitutional crises and dissolutions of parliament.[23] In the United Arab Emirates, the state's ongoing security crackdown against both Shi'a and suspected members of the Muslim Brotherhood "has contributed to the construction of a 'them and us' mentality which never before existed" among the country's wealthy, close-knit, and comparatively homogenous citizenry.[24] And in Bahrain, of course, the Shi'a- and secular-led opposition has sustained a decade-long protest movement, not on a platform of socioeconomic equality but in demand of *political* equality and other basic democratic reforms promised by King Hamad upon his succession. In the aftermath of the February 2011 uprising, the state has used the specter of Iranian expansionism to mobilize Sunni citizens against the opposition as an imagined Shi'a fifth column, polarizing society along sectarian lines and so hampering resolution of the political stalemate.

At the most basic theoretical level, then, this book aims to resolve this problem of the failed rentier state: a country swimming in historical amounts of oil and gas royalties yet unable to buy political disinterest. The key to explaining the puzzle lies in a factor omitted altogether from the standard theoretical framework: the region's structural tendency toward ascriptive group politics, a process encouraged both by the unique economic and political institutions of the Arab Gulf states as well as by interest-maximizing ruling families themselves. Whereas the allocative nature of the rentier state is thought to preclude the emergence of class politics, nevertheless, descent-based political coalitions can and do provide a viable focal point for mass political coordination in Gulf societies. This alternative avenue for political group formation not only leads to higher levels of popular political activity and dissent than the classic rentier model would predict, but also it is the institutional prerequisite behind another of the region's most salient political characteristics: the prevalence of "sectarianism," "ethnic rivalry," and other types of group conflict usually attributed tautologically to innate enmities, solidarities, and other ill-defined emotions.[25]

Citizens of rentier societies are said by the very nature of the economy to face strong incentives to compete independently for a greater personal share of state-distributed benefits. Yet, on the other hand, competing institutional forces push Gulf citizens in the opposite direction, toward a group-based politics to rival individual jockeying as the dominant political *modus operandi* in the rentier state. A closed political environment in which others' policy views are not easily identifiable, combined with an economic landscape that promotes individual rather than collective pursuit of material resources, works to privilege political coordination on the basis of ascriptive social groupings rather than cross-societal programmatic coalitions. Under such conditions, outward markers that communicate group descent—language, skin color, family name, and so on—transmit data not only about an individual's *social* affiliation, but moreover about one's likely *political* orientation as a member of a specific region, tribe, or religious or ethnic community.

Beyond facilitating political coordination among individuals, this process also allows Gulf leaders seeking to maximize their own shares of directly accruing external rents to capitalize on what Daniel Corstange calls an "ethnic subsidy": the cheap support of co-ethnics (or co-sectarians, fellow tribesmen, etc.).[26] As Corstange explains, co-ethnic support is relatively inexpensive in the sense that it is more easily inferred by ruling elites, who must distinguish between political supporters and opponents so that scarce resources are employed in rewarding the former and not dissipated on the latter. The upshot is a structural tendency toward political coalitions based on ascriptive descent as well as popular co-ethnic (or co-sectarian or co-tribal) support for governments.

Differences in resource endowments aside, this account helps to explain the relatively greater success of the more ethno-religiously homogeneous[27] Gulf states of Qatar and the United Arab Emirates in converting oil and gas revenues into stability. But it also suggests a reason why heightened political consciousness in Bahrain, Saudi Arabia, and Kuwait, while perhaps a headache for their respective ruling families, has not spelled disaster either. For while ascriptive social categories may come to supersede the individual as the basis of political action, these blocs do not emerge solely as contenders to the ruling family's political power but as competitors against each other—competitors certainly over their relative shares of state benefits, but so too over the very character of the nation itself.

Thus it happens that political involvement in the Gulf rentier state is not limited by the acquisition of material goods but is influenced crucially by the pursuit of intangible goods tied to one's group: its relative status in society, its relative political power as enshrined in state institutions, and its relative access to the ruling elite. No little energy is spent vying for greater allocation of resources and societal influence for one's group, decrying the inequitable distribution thereof and vilifying the opposing faction; yet in the end, a great deal of this effort is thereby directed not at the government but at the rival camp. In this sense a potentially destabilizing force can be captured or deflected by the state, which therefore has a direct interest—and most often a direct hand—in fostering and perpetuating these between-group struggles.[28]

Group Politics and the Rentier State

"Consider," Yates says of the rentier economy, "the following options for class-based politics: a declining rural-agricultural sector; a state-sponsored industrial sector; a booming service sector. Whence the revolution?"[29] Indeed, beginning with the earliest statements of the rentier thesis, theorists have held out little hope that anything resembling a traditional party system or cross-societal political movement might take root in an allocative society. Not only, as Yates laments, is there no natural social grouping like a taxpaying middle class whence such a push might originate, but moreover the patronage system itself incentivizes individual—not group—efforts to secure material benefits. "To the individual who feels his benefits are not enough," Luciani explains,

> the solution of manoeuvring for personal advantage within the existing setup is always superior to seeking an alliance with others in similar conditions. In the end, there is always little or no objective ground to claim that one should get more of the benefits, since his contribution is generally dispensable anyhow.[30]

So it is that in rentier societies the economy is assumed to offer little basis for political coordination, since "the politics of allocation states leave little ground for economic interests of citizens not belonging to the elite to be represented."[31] All this leads Luciani to make in passing what two decades later can only appear a rather prescient prediction, that in rent-based societies "parties will develop only to represent cultural or ideological orientations. In practice, Islamic fundamentalism appears to be the only rallying point around which something approaching a party can form in the Arab allocation states."[32]

The experience of the two Gulf countries that permit formal political groupings[33]—Bahrain and Kuwait—would seem to support this notion. There, the largest and most influential coalitions have formed along ascriptive social categories: religious ideology and denomination, and tribal versus nontribal background. Prior to a dramatic December 2012 electoral boycott in protest against changes to the country's voting districts, parliamentary politics in Kuwait was dominated by a large Sunni tribal coalition along with multiple blocs appealing to Shi'i and Sunni Islamists. The former are represented by separate coalitions for followers of the mainstream and Shirazi schools of Shi'ism, respectively, while Sunnis are represented by separate coalitions for followers of the Muslim Brotherhood and Salafist currents. Only two influential blocs run on programmatic platforms unrelated to social grouping: one made up of liberal and left-leaning intellectuals, and the populist Popular Action Bloc. Yet even these have felt the pull of sectarian politics. In February 2008, two Shiʿi MPs from the Popular Action Bloc were expelled from the group after attending a rally mourning the assassination of a top Hizballah commander.[34] They would later regain their seats as members of the main Shiʿa party, the National Islamic Alliance.[35]

Even more stark is the configuration in Bahrain, where since its reinstatement in 2002 the elected *majlis al-nuwwāb* has served primarily as a battleground for Sunni and Shiʿi societies. After ending its parliamentary boycott in 2006, the main Shiʿa bloc, the al-Wifaq National Islamic Society (al-Wifaq), faced consistent and intransigent opposition by tribal and Islamist Sunnis, whose primary preoccupation remained the obstruction of al-Wifaq's already limited legislative agenda. As in Kuwait, one Sunni society (al-Asalah) represents Salafis, and another (al-Manbar al-Islami) followers of the Muslim Brotherhood; while members of a third major bloc, known euphemistically as "independents," hail from influential tribes and families traditionally aligned with the ruling Al Khalifa. Such indeed is the extent of Bahrain's sectarian political alignment that when the results of the first fully contested election were announced in December 2006, even the government-affiliated *Akhbar Al-Khaleej* (*Gulf News*) could not avoid the conclusion that the politics of religion had won the day. Its lead headline summed up the vote: "After Announcement of the Final Results: Religious Control over the Council."[36]

Bahrain's political landscape in the post-uprising period is, if possible, even more entrenched along confessional lines, an outcome influenced, as before the unrest, by the state's targeted persecution of groups and individuals who dared advocate cross-sectarian political cooperation.[37] Since resigning from parliament en masse in February 2011 over the state's deadly response to mass protests, al-Wifaq has, despite government courting, reverted to its original position of legislative boycott, refusing to participate in September 2011 by-elections to refill the 18 (of 40 total) vacated seats, and again in regular elections held in November 2014. Other Shiʿa movements—both underground factions that splintered from al-Wifaq following its 2006 decision to enter formal politics,[38] as well as once-registered groups such as the Shirazi Islamic Action Society—have been effectively immobilized following the imprisonment of their political and clerical leadership and sustained persecution of their supporters. In the latter case, the society was dissolved altogether. In their place has emerged a heterogeneous, village-based Shiʿa street movement whose followers continue to engage in violent confrontations with security forces. A state-sponsored effort in 2011 to organize a pro-government Shiʿa society headed by a well-known (if unpopular) cleric gained more laughs than political traction.[39]

The Sunni community has witnessed similar political reconfiguration. Frightened by the magnitude and pace of Shiʿa-led protests in February 2011, behind which they saw the hand of Iran, Bahraini Sunnis quickly organized their own mass demonstrations in support of (and at least partly at the instigation of) the government. Headquartered appropriately at the state mosque, this Sunni coun-termobilization largely sidestepped the existing formal Islamist societies to assume a populist character, being led by previously-obscure figures closer to academia than politics. From this Gathering of National Unity (TGONU) would eventu-ally split other groups, most notably the youth-oriented Sahwat al-Fatih associ-ated with the Muslim Brotherhood. With the state's forcible end to mass protests, however, the negative impetus binding Bahrain's otherwise dissonant Sunni community has gradually faded, and with it most of the momentum of the new movements.

One sees, then, that in the contemporary Gulf context ethno-religious categories—confessional and even sub-confessional membership—offer the most viable focal points for political coordination in a type of society that otherwise, for lack of both political and economic institutions to help channel citizens' in-terests, presents significant organizational barriers to mass cooperation. While identification of an individual's religious tradition may seem a crude substitute for knowledge of his actual political preferences, given the region's relatively bar-ren political landscape deficient of institutions such as nongovernmental organi-zations, independent media, and proper political parties able to provide informa-tion about others' political characteristics—in the absence of such proxies, one relies in one's choice of prospective political allies instead on the only data avail-able: names and genealogies, language and accent, skin color, geographical ori-gin, and so on.[40] In short, politically minded individuals must depend almost exclusively upon ascriptive social categories, whether (according to geographi-cal and demographic accident) ethnicity, family or tribal descent, or religious denomination.

While the inferences gleaned from such external cues will likely only approx-imate the true natures of fellow citizens, they are, first of all, very simple and cheap to observe. Given the impermeability of ethnic and, to a lesser extent, religious boundaries, moreover, they are also likely to remain quite accurate even over time. Political coordination in this way occurs most likely among individuals of simi-lar ascriptive makeup, who form a common bond that may have a dubious basis in historical fact or in actual shared political interests,[41] but one that "denotes not just a certain stream of belief but a certain version of peoplehood."[42] Though they may be accentuated or muted by various means, deliberate or accidental, in the Gulf context ethnic and religious categories are natural foci for political coordi-nation: they are inherently *political* categories.[43]

Importantly, however, the same key structural variables that underlie this ten-dency toward ascriptive group coordination also help explain why some ascrip-tive categories achieve political salience while others do not. Corstange explains that it is the combination of high membership observability and low boundary permeability, rather than anything intrinsic to these categories themselves, that privileges the formation of ascriptive coalitions in low-information political environments. These factors "[reduce] compositional ambiguity, both at a given point and over time, and so help category members identify the category itself as well as one another, making their common interests *common knowledge*."[44] That one's affiliation as a Sunni or Shi'i bears political significance in Bahrain depends, then, in the first place, on the fact that members of both communities can identify each other readily and with a high degree of certainty.

By contrast, while most Bahrainis can probably name individuals and families known to belong to less populous ethno-religious subgroups such as Persian Shi'a (the 'Ajam[45]) and the Sunni Huwala,[46] and while members may be relatively well known to each other, still these categories involve a substantially higher barrier to both in-group and out-group identification as compared to simple sectarian affiliation, which can be inferred from a diverse set of proxies including even outward cues such as dress, physiological features, and especially language. They are, in addition, relatively more fluid than the question of religious denomination, which is reinforced through daily practice. Whatever the other factors that have worked to limit the political salience of these two subgroup identities in Bahrain, not least important is this relative ambiguity in group composition both in general and over time.

Finally, beyond its role as organizational focal point, there is another manner in which religion in particular may promote political coordination, namely when the doctrine itself carries (or can be interpreted as carrying) lessons or prescriptions informing the political behavior and principles of its followers. These may exist, for example, in the form of positive or negative regulations regarding proper actions or values in the political sphere, or, alternatively, they may arise from the very historical events and circumstances surrounding a religion's genesis and development. Such history bears special political relevance for schismatic traditions, which by definition exist in contraposition to some greater spiritual and so political authority. In the case of Bahraini Shi'a, of course, this status as outsider is reinforced by the group's historical social and political marginalization.[47]

The personalities and events of these founding days in particular can evoke powerful remembrances of political grievance when institutionalized in ritual and lore—and when put to good use by shrewd political entrepreneurs looking to rally the troops. The most poignant example of this is the Shi'i ritual of Ashura, an annual ceremony that culminates in a frenzy of mass self-mutilation in mourning of the murder of the (Shi'i) Imam Husain ibn 'Ali at the hand of the (Sunni) Umayyad caliph Yazid I, to whose rule the Imam refused to pledge political allegiance. These days of heightened religious emotion usually correspond to increased sectarian tensions and, in the case most visibly of those Shi'a making pilgrimage to holy sites in Iraq, violence and bloodshed.

Religion's role in augmenting political interest and activity is not limited to Shi'a or other ascriptive-cum-political out-groups, however. By shaping political orientations and spurring activism among the latter, it works simultaneously to marshal regime allies in defense of the status quo and, more to the point, in defense of their favorable cut of the political-economic pie. While defenders of the state may dismiss the critics as blaming their own economic and political failures on invented discrimination, inevitably they cannot but be drawn into the group dichotomy injected into the national consciousness, and, aided by parallel sectarian tensions at the regional level, they come to define themselves along the same dividing line.[48] So it is that religious identity comes to play a dual role in the polity: for members of both the favored and the disfavored constituency, greater ingroup identification stimulates increased political action and alters political opinion; but among the former these actions and opinions are undertaken in defense of the government and the prevailing system that it safeguards, while among the latter they are expressed in opposition.

In addition to the institutional factors contributing to the emergence of group-based politics in the Arab Gulf states, there remains of course the role of governments—of ruling families—themselves, whose overarching concern for regime security is often served, at least in the immediate term, by the deliberate enhancement, even cultivation, of sectarian, tribal, and other latent social divisions. With few exceptions, the formal institutions and national narratives developed by Gulf states have been designed not to de-emphasize the political salience of religious and other ascriptive distinctions, but precisely the reverse: to enhance society's latent potential for group-based competition. Electoral rules and voting districts are manipulated to foster intergroup contestation, while selective employment and naturalization policies construct tiers of citizenship ascending toward an ideal type. Glorified in official histories, these single visions of peoplehood neglect competing group identities and experiences, in some cases—such as in Bahrain— even those shared by a majority. To understand states' incentive to promote such seemingly deleterious conflict among citizens, one must probe more deeply into the received wisdom of the rentier bargain, rethinking the facile and (from the standpoint of ruling families) overly generous postulate of material benefits in return for political loyalty.

Strategies of Rentier Rule

It has been said that the contemporary record of Gulf politics would seem to demand a basic reevaluation of the implicit "social contract" presumed to exist between rulers and ruled. For not only have the Arab Gulf states failed to purchase political autonomy from their citizenries, but, empirical discrepancies aside, it is clear that such an open-ended wealth-for-silence agreement never operated in the first place. Rather than deploy limited resources inefficiently upon the whole of society, the controllers of rentier states instead seek to maximize their own consumption of the material benefits of rulership, by offering citizens only the minimum allocation necessary to ensure a winning coalition of supporters.[49] Indeed, to what purpose power if one must trade away its earthly rewards? To recognize this incentive, on clear display in the colossal (and secret) discretionary budgets of Gulf ruling families, is to recognize that the primary task of rentier governments is not simply the distribution of resource wealth to the population, but doing so as cheaply as possible.

Accordingly, the most elementary question concerning the rulers of rentier societies is how to achieve the optimal balance between economic and political autonomy, that is, how to maximize simultaneously both (a) private enjoyment and discretionary employment of directly accruing resource wealth, and (b) freedom from popular accountability through economic appeasement via distribution. Err on the side of the former and one risks losing the support of society; on the side of the latter and one both squanders scarce resources and jeopardizes the support of one's family, whose members also expect to be rewarded.[50] This implicit political negotiation is thus not unlike the "ultimatum game" of experimental economics, wherein one player offers a portion of a divisible prize to a second, who can accept or reject the proposed allocation. If the offer is accepted, both players are better off, even if (depending on the proposal) one relatively more so than the other. But if the allocation is rejected as unfair, neither receives anything.

The possible solutions to this rentier dilemma are various and can be grouped into at least four distinct strategies. A first, which one might call *liberality*, would consciously overpay society at the expense of elite consumption, sacrificing a portion of rulers' private enjoyment of rents in order to ensure widespread popular support and political autonomy. A second strategy, *economic diversification*, would seek to reduce the burden of distribution by augmenting or replacing state-provided benefits with benefits provided by private firms. A third, *political diversification*, would also seek to reduce dependence upon distribution, yet not through privatization but by expanding and deepening the bases of political legitimacy away from simple material benefaction. This strategy would seek to move beyond the traditional rentier citizen-state relationship, augmenting or replacing economic benefits with intangible goods such as the safeguarding of local culture and religion,

Economic autonomy	**Political autonomy**
Enjoyment of rents	Distribution of rents

Figure 1.3. The rentier trade-off.

knowledge and education, and political *stability* over against political accountability. A final strategy, which one might call *political segmentation*, would discriminate a country's political markets, disproportionately rewarding a class of citizen supporters and disproportionately excluding the remainder from the rentier benefits of citizenship. Here material benefits are not dissipated across the whole of society but concentrated on a finite constituency whose support is sufficient to ensure the continuity of the regime.

Of course, the appropriateness and effectiveness of each strategy will vary across societies, being largely dependent upon structural variables such as the magnitude of external rents; the size, demographic homogeneity, and geographical dispersion of the citizen population;[51] and the degree of unity among the ruling elite. A strategy of liberality, for example, requires a level of resources out of reach even to most rentier states. As reported already, Saudi Arabia's resource revenues, although vast in raw terms, amounted to only around $16,400 per citizen in 2013, Oman's only slightly more at $18,300, and Bahrain's $29,000, whereas Kuwait, the UAE, and Qatar earned $73,000, $131,500, and $428,000 per citizen, respectively. Even with outside aid, such as the GCC provided Bahrain and Oman in response to unrest in 2011, the former category of states simply lacks the funds to buy widespread political support through direct distribution. A country like Qatar, on the other hand, with some $107 billion in oil and gas royalties to be shared among a mere 275,000 nationals, may offer a quite generous and relatively egalitarian distribution of citizen benefits without bankrupting the state or limiting the discretionary spending of the ruling elite. The latter condition is not to be overlooked, since a policy of liberality, ceteris paribus, should tend to increase the likelihood of internal dissent among members of the ruling family, who may not share a leader's risk aversion or economic moderation.

Political segmentation entails a different set of dangers. Although this strategy minimizes the cost of distribution and cultivates a core constituency of supporters invested in the politico-economic status quo, the systematic differentiation of citizens requires a precarious political balancing act susceptible to precisely that outcome meant to be avoided in the first place, namely political dissent and instability. At the same time that the state garners the support of one class of citizen, most often defined by shared ascriptive descent, it earns the dissatisfaction and enmity of many others, in addition to a society polarized along group lines. A state may overestimate the relative strength of its core constituency or underestimate the power of those disproportionately excluded. Moreover, even if the latter lack the numbers or arms to pose an existential threat, they may serve to hamper economic productivity through debilitating protest action, heighten regional tensions, or damage a state's reputation.

Diversification, finally, while popular in principle, has proven difficult to achieve in practice. Certainly much of this, particularly on the economic side, owes to the inherent difficulty of altering deeply entrenched socioeconomic structures and incentives: dependence upon migrant labor, public-private sector wage and productivity imbalances, low female workforce participation, and so on. Yet equally intractable has been the problem of retrenchment, the relaxation of policy and re-shuffling of priorities in times of perceived political vulnerability. In the wake of the February 14th uprising, for instance, Bahrain offered in lieu of political change promises of new economic benefits meant to appease both ordinary citizens and elites. For the former, it announced a generous social welfare package including increased salaries, a cost-of-living stipend, and plans for new subsidized housing. For the latter, it suspended an innovative yet (among business owners) highly unpopular tax on foreign labor meant to incentivize the hiring of citizens and so reduce dependence on non-nationals.[52] Other Gulf governments would follow suit, undermining longstanding efforts to promote national employment in the private sector.[53] Even Qatar, the one country spared by the Arab uprisings, opted for short-term political expediency over long-term economic sustainability. In September 2011, absent any discernible popular pressure, the state unveiled a preemptive 60 percent increase in salary for Qataris working in the public sector, doubled to 120 percent for those in the police and military.

In practice, however, these four rentier strategies are rarely pursued in isolation, and several are inherently complementary. The formidable task of economic and/or political diversification, for example, may be pursued most seriously and confidently when it exists as a complement—as a fallback—to the more proven strategy of liberality. For while the potential benefits of diversification are high, so too are the financial costs and political risks. Economic diversification entails, *inter alia*, considerable investment in education as well as possible capital flight as a result of unwelcome labor market regulations (for instance, quotas for nationals in the private sector). Efforts at political diversification, on the other hand, might be associated with even more spending, particularly on physical infrastructure: on mosques, museums, monuments, malls, traditional *suqs*, and even larger cultural-economic mega-projects such as Qatar's Katara and Education City, Saudi Arabia's planned King Abdullah Economic City, and the UAE's planned Dubailand, Masdar City, and Mohammad Bin Rashid City.[54]

Yet this investment in long-term political stability (and to a lesser extent economic stimulus and diversification) must come at the cost of short-term stability, inasmuch as these same funds could have been distributed directly to citizens with some marginal political gain. For states that enjoy the resources to be generous to individual citizens even as they spend billions of dollars on legitimacy-enhancing initiatives, this opportunity cost may be small or imperceptible. But for poorer states, the risk that public goods will not be accepted in lieu of private benefits is not trivial. Hence, experimentation with alternative economic and political models—whether weaning citizens off public-sector life-support or winning their loyalty through appeals to culture, religion, and other intangible goods—is most likely among those rentier states that already are able to appease most citizens through liberal distribution.

Similarly, a strategy of segmentation, in structuring society into rentier winners and losers, naturally promotes the cause of political diversification. By engineering the rise of systematic dissatisfaction and controlled, group-based opposition, the state renders itself not only economically indispensible to its supporters, but also politically indispensable in the face of an organized minority—or, in a case like Bahrain, majority—of disenfranchised citizens eager to rewrite the entire system. The state, in other words, becomes guarantor not simply of the politico-economic status quo, but of an entire faction of society against its real or engineered rivals. Under such circumstances, fear may come to displace distribution as the primary mechanism bonding citizen to state.

The nature and strength of the individual-level link between economic satisfaction and political quiescence in rent-based economies will thus depend necessarily on the strategy of rule adopted by a state. If the relationship between citizen and government is rooted wholly in economic patronage, and if the latter is extended universally, then the case may be similar to that described in the foundational rentier state literature, with citizens' political orientations and behavior determined primarily by their relative satisfaction with material benefits. But if economic distribution is only a part of a state's wider political legitimization strategy, or is limited disproportionately to a certain subset of the population, then one should have different theoretical expectations.

In the former case of diversification, one should expect material satisfaction to compete with other, intangible factors in predicting citizens' political views and actions. In the case of segmentation, one should expect the individual-level relationship between economic and political satisfaction to operate among some citizens—that is, members of the economic-cum-political in-group—but weakly or not at all among others. Alternatively, it is also possible that the political effects of segmentation come to override the economic; that members of the in-group become linked to the state not as economic patron but above all as protector of the political status quo and of the prevailing balance of power among social factions. This would seem to describe the present situation of the Sunni community in Bahrain and to a lesser extent that in Saudi Arabia, though this notion of the state as veritable "protection-racket" applies more widely to the Arab world.[55]

Group Politics and the Limits of Rentierism

So far we have seen how, in the barren landscape of the Gulf, lacking both an economic basis for mass coordination as well as institutions to channel group interests, the process of political coalition-building favors alliances based on outwardly observable, ascriptive social categories. As a result, political cooperation becomes most likely among citizens of similar ethno-religious makeup, and, in distributive states with diverse and/or regionally diffuse populations, rulers face strong economic incentives to segment their political markets according to these same descent-based groupings. Not only is there no universal rentier bargain tying all citizens to rulers of distributive states, therefore, but also the lines separating those who are party to the agreement from those who are disqualified are not drawn arbitrarily.

As such, one can readily perceive how the group-based political mobilization witnessed in Bahrain and elsewhere across the Gulf is not easily suppressed, that is, why those who incline toward political involvement are not easily co-opted using the typical pressure-relieving mechanisms thought to be available to Gulf regimes *qua* allocative economies. The latter, one will recall, are said to be able to convert external rents into political quiescence via at least two distributional mechanisms—guaranteed employment and non-taxation—and arguably a third avenue: physical repression. Yet the introduction of societal division along ascriptive group lines serves to handicap rentier states by, at best, making these options less efficient and, at worst, by taking them off the table altogether.

Bahrain has gone further than any other Gulf country in segmenting its political market, mainly on the basis of sectarian religious affiliation but to a lesser extent along the lines of tribal versus non-tribal.[56] Despite limited political liberalizations introduced upon the succession of King Hamad, the state has worked systematically to dilute the political influence of Shi'a nationals, which according to my Bahrain mass survey make up somewhere between 53 and 62 percent of the citizen population.[57] Bahrain's voting districts are gerrymandered utterly along Sunni-Shi'a lines, precluding an opposition majority in the elected but toothless parliament and limiting the electoral prospects of societies not based on sectarian affiliation. Ostensibly for fear of their ties to Iran and transnational religious movements, Shi'a are also disproportionately excluded from those ministries charged with the exercise of state power, and they are altogether disqualified from police and military service. This relative exclusion from government employment also limits access to other state benefits, such as public housing, priority for which is given to new foreign recruits for the army and security services, namely Sunnis from Pakistan, Yemen, Syria, Jordan, and elsewhere.

From the perspective of Bahrain's rulers, then, the state is struck in a veritable catch-22, wherein the very attempt to purchase political stability in fact serves only to open the door to increased instability. Specifically, the more Bahrain would seek to buy the political loyalty of opponents and would-be opponents using the most comprehensive clientelistic tool available to it *qua* rentier economy—private benefits conferred through employment in the public sector—the more it exposes itself to exactly that danger meant to be relieved in the first place, by inviting those citizens deemed most dangerous to walk in, so to speak, through the front door. As a result, government agencies deemed politically or militarily sensitive are made off limits to those outwardly identifiable as potential regime opponents, begetting a situation in which state employment is no longer an effective measure by which to procure political loyalty, but demonstrable political loyalty—in effect, the right family name—a prerequisite for most forms of state employment. This two-tiered system of rentier benefits, including police and armed forces that would prefer to employ Sunni non-nationals than take a chance with Bahraini Shiʻa subservient to their co-sectarians in Iran, works only to divide society further between those with a private stake in the state and those who feel not only unfairly excluded from it but indeed unwelcome in it.

What is more, in the aftermath of the Iranian Revolution and especially the post-2003 empowerment of Iraq's Shiʻa majority, Shiʻa populations across the Arab world are seen as increasingly forceful in their demands for greater authority, invoking historical claims of political right rooted in the very origins of Islam. In Bahrain, this perception, reinforced by two decades of organized Shiʻa opposition, serves to mobilize members of the state's Sunni support base. In order to offset a perceived growth in Shiʻa—and, by association, Iranian—influence, Sunnis organize themselves as a counterweight to perceived domestic Shiʻization and external interference, making in turn their own political demands upon the government. Indeed, a perceived leniency in dealing with opposition protesters led some security-minded Sunnis to openly challenge King Hamad himself amid continued unrest following the February 14th uprising. Popular interest in political participation becomes, then, not a function of material well-being as per rentier expectations, but one of religious identification and regional power struggles.

Accordingly, to the extent that economic satisfaction is a systematic determinant of political views and behavior in Bahrain, one should expect it to be so disproportionately among members of the Sunni in-group and, in any case, to be overshadowed by competing factors such as sectarian affiliation and orientation. Even among Sunnis, moreover, one might expect that the community's support for the state stems not primarily from the material rewards it delivers (or promises) but from the more crucial intangible benefit it provides: stability and security in the face of an emboldened domestic opposition with feared links to enemies abroad. As Justice Minister Khalid bin ʻAli Al Khalifa told *The Economist* during preelection turmoil in 2010, Bahrain's ruling family conceives itself " ʻa buffer zone' between Sunni and Shia."[58] This message is not delivered to foreign audiences only.

With such a surplus of political energy and limited means to diffuse it, it is thus little wonder that the more diverse societies of the Gulf—especially Bahrain, Kuwait, and Saudi Arabia—have been less successful than the other GCC countries in converting their resource windfalls into political stability. Indeed, one must be surprised that some, not least Bahrain with its full Shi'a majority, have faired as well as they have. Yet at the same time, in vying with each other over additional material allocation as well as over the more fundamental issue of relative group status in the polity, a great deal of citizens' energy is thus expended on horizontal contestation, a fight officiated by the ruling family as by a referee in a boxing match, with less directed vertically toward the official himself. This lack of cross-societal coordination helps explain why the Gulf's more heterogeneous citizenries, even if demonstrating higher levels of political activism than the rentier thesis would predict, still have not on that account posed an existential challenge to their rulers, who are, all things considered, little worse for the wear.

Summary

The foregoing represents a theory of group-based political coordination in the supposedly apolitical rentier societies of the Arab Gulf, one that dispels the myth of the politically agnostic oil sheikh as the region's archetypical citizen, and suggests a new variable to help account for observed differences in stability among the Gulf states themselves. It began by considering how far our inquiry might be informed by existing empirical tests of the rentier state framework, concluding that such studies suffer from basic methodological and theoretical limitations. Principal among those of the former category was the modern preoccupation with the question of democracy and democratization, a concern that has seen the monopolization of the rentier research agenda by works competing against each other to explain (or controvert) the apparent relative lack of democracy in resource-rich nations, with nearly all of these employing exactly the same dependent variable from exactly the same dataset: the Polity IV scale of regime type. Not only is this choice out of line theoretically with the rentier paradigm's concern for regime *stability* rather than regime *accountability*, but the resultant quantitative analyses are, as typically undertaken, susceptible to bias. In particular, when some continuous measure of rentierism is used to predict a country's regime score, the resulting negative and highly significant relationship is an artifact of a small number of outlying cases: the six GCC states along with Brunei and Libya. When these observations are removed, so too is the statistically significant link between resource rents and authoritarianism; all that remains is (an already obvious) such association among a finite group of states—the rentier states—and a compelling reason to refocus the analytical effort back toward the latter.

Yet even in the absence of these methodological issues, we continued, there remains still an underlying theoretical problem plaguing extant efforts to study the rentier state quantitatively, namely their failure to investigate the actual individual-level causal mechanisms that form the basis of the rentier hypothesis itself. If the theory posits that allocative states achieve stability by buying off citizens with rent-funded benefits, then evidence of a link between rent income and macro-level political outcomes like regime type, democratic transition, or even taxation and public spending rates is not proof of the theory itself; does not verify its internal causal story, but simply restates its own motivating observations. It is as if one were to suggest that a parliamentary candidate was elected by bribing prospective voters and offered as evidence the fact that he spent ten times the amount of his competitor: at the end of the day one would like to see his supporters' incriminating bank statements or, short of that, at least to ask them the reasons behind their votes. So too, for convincing proof of the rentier explanation one needs evidence that, among ordinary citizens of rent-based regimes, there exists a systematic relationship between material well-being and political behavior.

The remainder of the discussion elaborated several important theoretical reasons why one should *not* expect a universal association of this sort in the Arab Gulf context, and thus no inexorable link between oil and gas royalties and political acquiescence. In the first place, it was said, contrary to rentier assumptions, the region's unique political and economic institutions do not serve to preclude mass political coordination but rather to privilege a certain *type* of political cooperation, namely coordination on the basis of outwardly observable and relatively stable social categories such as ethnicity, religion, tribal or regional affiliation, and so on. The result is a structural tendency in the Gulf toward ascriptive group politics.

This institutional tendency, we continued, is reinforced by two additional factors. A first is the region's enduring politico-religious schism between Sunni and Shiʻi Islam. Beyond offering a viable basis for popular political mobilization among members of both communities, this conflict simultaneously works to hinder the normal function of the rentier state by precluding the most common avenues of political buy-off available to allocative governments. Because Shiʻa are today perceived by Sunni citizens and rulers as being not only religiously deviant but also politically subservient for doctrinal reasons to religious authorities in Iran, Iraq, or elsewhere, ruling families find it difficult to appease their discontented Shiʻa constituencies as they do not trust them enough to allow them to serve in leading public employers—the police, military, and power ministries. Moreover, because many Shiʻa believe they have a collective right to political authority based on memories of injustice and betrayal rooted in the very foundations of Islam, governments cannot easily pacify their political demands through economic bribery. At the same time, the prospect of an emboldened Shiʻa populace operating at the behest of a belligerent Iranian regime makes it intolerable for Sunni nationals to remain on the political sidelines, further undermining the myth of the apolitical rentier society.

Finally, it was said, the rulers of rentier states themselves have a direct economic and political interest in the promotion of intergroup contestation rather than cross-societal cooperation, whether between Sunna and Shi'a, tribal forma-tions, or other social groupings. Far from working to lessen such factional con-flict, then, Gulf ruling families have devised various ways of magnifying and institutionalizing group difference, including via citizenship and electoral laws, formal representative institutions, and exclusionary national narratives. Rather than encourage cross-cutting citizen coalitions, electoral rules and procedures are manipulated to ensure the continued political salience of ascriptive distinc-tions, especially those based on tribal and confessional affiliation. Rather than unify citizens around a common vision of peoplehood with which all citizens can identify, most national identities and histories are crafted in the image of ruling elites. As a result, not only are the experiences and heritage of some or even most citizens, including Shi'a and nontribal Sunnis, not represented in of-ficial state doctrine, but also these citizens lack the genealogical background necessary to share in traditional expressions of support for political elites as co-religionists or as leaders of an extended national tribe. In their place, ex-cluded communities maintain competing identities and traditions, preserved in histories emphasizing injustice and marginalization that, in the case of sectarian relations, draw on Islam's centuries-old conflict over political succession and leadership.[59]

The outcome of these conceptual revisions to the rentier framework is a changed set of predictions about the individual-level relationship between eco-nomic and political satisfaction supposed to obtain among Arab Gulf citizens. In countries home to sectarian and other types of group politics, such as describes not only Bahrain but most of the GCC, one should expect that popular interest and participation in politics should be a function not only of material circum-stances, but also of group affiliation and the personal political salience of one's group identity. Among members of the political out-group in particular, one should expect that political opinions and actions are impacted relatively little by varia-tion in economic satisfaction, since members are not party to the implicit rentier agreement operating between rulers and their (usually co-ethnic) support base to begin with.

Regarding the determinants of political orientation among the latter group of state supporters, one might formulate competing hypotheses. On the one hand, since these citizens *are* systematically linked to the regime through patronage, one might expect that their support should indeed be determined above all by rela-tive satisfaction with their economic situation, that is, by the degree to which the state is holding up to its half of the rentier bargain. Yet insofar as such a state also provides the vital *intangible* benefit of "protection" against members of the excluded community who would seek fundamental change of the system, it might be that recognition of this valuable service comes to override strict economic concerns among members of the political in-group.

Group Conflict in the Rentier State: The Case of Sunni-Shi'i Relations in Bahrain

With the ongoing, if slow, extension of social science survey research to the Arab Gulf, the foregoing represents not simply a theoretical critique, but specific propositions that can be interrogated empirically and at the appropriate level of analysis using survey data. The modified framework outlined above will in the ensuing chapters undergo empirical evaluation with reference to Bahrain in particular. This case offers important practical and theoretical advantages. In the first place, due to the sensitive status of politics, religion, and sectarian relations in Bahrain as everywhere in the Arab Gulf, few have succeeded in administering nationally representative surveys in the region that ask the sort of questions one requires in order to analyze the link hypothesized here between economic welfare, group identity, and political orientations.

It is telling that in its twenty-year history, even the near-comprehensive World Values Survey, which has been fielded in more than 150 countries since 1981 and whose questions are not particularly sensitive, has been administered in the Arab Gulf, with one exception, only within the past five years. Moreover, these surveys, undertaken in Saudi Arabia (2003), Qatar (2010), and Kuwait (2014), were for the most part unable to field the most revealing political questions about normative opinion toward the state and participation in various types of political activities. Similarly, excluding my Bahrain survey, only one other GCC state counts among the seventeen countries included in the first two waves of the regional Arab Barometer survey project. This 2011 survey of Saudi Arabia represents only 1,405 (approximately 7 percent) of 20,890 total interviews conducted. The first practical difficulty, then, is the sheer lack of scientific and nationally representative survey data being collected in the Arab Gulf.

The second is that the data that are available lack ethnic, confessional, and other group identifiers for respondents, meaning that one cannot distinguish, say, Sunni from Shi'i. This is of course not by chance, as questions about sectarian affiliation touch on sensitivities of both respondents and governments. But to test a theory that predicts intergroup variation in responses, it is precisely this information that one requires, rendering extant Arab Barometer and WVS data of limited use. Hence, the real practical advantage presented by the case of Bahrain is that, owing to a nationally representative survey of 500 randomly selected households I administered in early 2009, the requisite data actually exist. A full methodological overview follows in chapter 4; it is enough to say here that Bahraini interviewers administered the standard Arab Barometer survey instrument and, aided by the plain linguistic and geographical segregation of the country's Sunni and Shi'i populations, were able easily to infer the confessional affiliation of respondents. The survey, then, was to the author's knowledge the first (non-state) effort to collect individual-level denominational data in the Arab Gulf.

The impetus for and theoretical importance of the Bahraini case lies not in a lack of previous data, however, but in the character of Sunni-Shiʻi relations in this rentier state. Almost since the day the ruling Al Khalifa dynasty arrived in Bahrain in 1783 along with its Sunni tribal allies, having just wrested the island from the Persian Empire, its relations with the indigenous Shiʻa inhabitants have been marked by economic exploitation, social detachment, and, often, political conflict. This is especially true of the period following the Iranian Revolution and later the U.S.-led regime change in Iraq. The prospect of Shiʻa populism inching its way toward the Arab Gulf states set off a destructive geopolitical cycle from which neither Bahrain nor the region has yet recovered.

In 1981, Bahrain thwarted an alleged coup attempt by the Iranian-linked Islamic Front for the Liberation of Bahrain. Redoubled political repression helped precipitate a mass Shiʻa uprising that would engulf the second half of the 1990s, subsiding only with the sudden death of Emir ʻIsa bin Salman in March 1999 and an immediate pledge of political liberalization by his son and new ruler Hamad. But it soon became clear that the heralded reform project would fall far short of popular expectations, and this just as Iraq's longsuffering Shiʻa majority was poised to grab the reins of power. Frustration again boiled over, mobilizing Shiʻa and secular reformists, alarming the regime's traditional Sunni support base, and affording additional fuel for a deliberate state strategy of divide and rule conceived by more security-minded members of the ruling family. As we shall observe in the next chapter, these two overlapping conflicts[60]—citizen-state and citizen-citizen—continue not only to complicate Bahrain's task of buying political calm, but also to ensure that this calm, once broken, is not easily restored.

2 Al-Fātiḥ wa al-Maftūḥ
The Case of Sunni-Shiʻi
Relations in Bahrain

Tɪɴʏ ᴛʜᴏᴜɢʜ ɪᴛ is, the 33-island archipelago of Bahrain, situated 15 miles off the eastern coast of Saudi Arabia in the Persian Gulf, is an ideal location in which to examine the disruptive influence of group-based political mobilization on the normal function of the rentier state. Indeed, for a kingdom but half the size of London, Bahrain holds a number of distinctions: the global center of pearl production and trading until the 1930s; the first Gulf country in which oil was discovered and mined; the former home of colonial Britain's Residency of the Persian Gulf and present base of the U.S. Fifth Fleet; and, since the 2003 fall of Iraq's Baʻathist regime, the only Middle East nation still ruled by a Sunni minority. While the exact proportion is itself a much-debated and highly divisive issue, it is generally agreed that, despite a decade-long campaign of naturalizing Sunni foreigners, Shiʻa still comprise somewhere between 55 percent and 65 percent of the total population of Bahrain, making it one of just three Middle East states, along with Iran and Iraq, wherein this perennial minority holds an absolute majority.[1]

That Sunnis are here outnumbered, however, is not what makes Bahrain interesting to study. More important is that relations between Bahraini citizens and the ruling Al Khalifa tribe *qua* government, to a degree unparalleled anywhere else in the Arab Gulf, fail to operate according to the standard patron-client formula represented by the rentier state model. The lesson to be learned of the past two decades of political turmoil in Bahrain—to say nothing of the popular uprising and sectarian clashes of the 1950s; the showdown over a 1965 Public Security Law and ensuing dissolution of parliament in 1975; and the Iranian-linked failed coup attempt of 1981—is that either Bahrain's rulers are singularly inept at using the country's sizable oil windfall to placate would-be opponents, or there is something about the way politics operates in Bahrain that renders the state systematically unable to do so.

The "Opening" of Bahrain: The Enduring Legacy
of the Al Khalifa Conquest

The Al Khalifa's 1783 capture of Bahrain from the Safavid Persian Empire is immortalized for all who see the island in the ubiquitous references to "the conqueror" (*al-fātiḥ*) himself, Ahmad bin Muhammad Al Khalifa. Having crossed the bridge into Manama from the airport in Muharraq, one likely turns south onto the Al-Fatih Highway, passing on the way the enormous Al-Fatih Grand Mosque, by far the largest place of worship in Bahrain and one of the largest in the Islamic world. The mosque is flanked to its west by the Gudaibiyya Royal Palace and to its north by a newly opened National Library. More than just a painful reminder of the social and political upheaval occasioned by the Al Khalifa's arrival, however, the prominent place of "Al-Fatih" in the national lore and present-day geography of Bahrain represents for the country's Shiʿa population something more hateful. For while the term (literally, "the opener") can mean "the conqueror" or "the victor" in the military sense, it also carries overt religious overtones that certainly are not lost on ordinary Bahrainis as they would not be on any Arabic-speaking Muslim.

When seventh-century Muslim armies fought to spread their nascent religion across the Arab world and beyond, they were said to be effecting "*fath al-islām*"—the "opening of Islam"—a euphemism for the conversion and, upon refusal, subjugation of non-Muslim peoples.[2] Its use in the Bahraini context, then, implies not simply that the island was conquered militarily by Ahmad Al Khalifa and his Sunni tribal allies, but that it was "opened" for Islam—that is, for the true Islam—in view of its indigenous Shiʿa inhabitants and its prior status as a protectorate of Safavid Persia, which since 1501 had embraced Shiʿism as a state religion.

The continued glorification of this event and of this terminology on the part of Bahrain's rulers is but one feature of a larger national identity crafted in the image of the ruling dynasty. Distinguishing as it does those citizens who can and cannot claim to possess the hereditary qualifications to be "true" Bahrainis, this exclusionary narrative aptly symbolizes the elementary sociopolitical divide separating ordinary Sunni and Shiʿi citizens. As pithily expressed by popular Shiʿi cleric Sh. ʿAbd al-Wahhab Husain, a powerful force in the uprisings of the 1990s and February 2011 (and currently serving a life sentence for his role in the latter), the difference between Sunna and Shiʿa in Bahrain is the difference between "*al-fātiḥ wa al-maftūḥ*": "the opener and the opened."[3]

Precisely how Bahraini society looked prior to the arrival of the Al Khalifa and their tribal allies is the subject of much speculation. For their part, the Shiʿa make, as Khuri notes in his yet unrivaled sociological survey of Bahrain, an unlikely religious interpretation:

> They say that Bahrain had three hundred villages and thirty cities and towns before the Al-Khalifa conquest, each ruled by a jurist who was well versed in Shiʿa law. These three hundred and thirty jurists were organized into a hierarchy headed by a council of three, elected by an assembly of thirty-three who, in turn, were acclaimed to power by the jurists of the whole country.[4]

This fanciful portrait of pre-Al Khalifa Bahrain governed by the magical number three is but one element of what Louër has called a "myth of golden age," a tale ingrained into the collective consciousness of the region's Shi'a that draws upon the historical usage of the appellation "Bahrain" to refer to all the Gulf coast from Basra to the Qatar peninsula, the heart of this ancient territory being the modern Bahrain archipelago along with the oases of al-Qatif and al-Hasa now part of Saudi Arabia's Eastern Province.[5] This narrative, recounted to her in interviews by Saudi and Bahraini Shi'a alike, Louër summarizes as follows:

> There was a time when the Shias of Eastern Arabia were united in one single country called Bahrain extending from Basra to Oman. Its inhabitants were called the Baharna and had embraced Shiism since the beginning of Islam. Bahrain was a wealthy country blessed by several natural resources: fresh springs, arable lands and pearls. People were living a simple but fully satisfactory peasant life in accordance with the prescriptions of the Imams. It was a time of social harmony and order. Everything changed when the Sunni tribes—the Al-Khalifa and the Al-Sa'ud—took over the region, appropriated the natural resources for their own use and imposed their brutal and autocratic manners on the native population. They not only oppressed the Shias but cut their unity by breaking the organic ties between the islands and the inland. Since then, marginalized Shias have fought to recover their legitimate rights as the native inhabitants of Ancient Bahrain.

These tales of the glory days of Bahrain, of the time it was ruled by enlightened Shi'a jurisprudents for the sake of its Shi'a inhabitants and unspoiled by unjust alien intrusion, are not mere bedtime stories. Historically embellished and selective though they are—where, for example, are the Portuguese and the Ottomans?, who by 1550 had already divided "Ancient Bahrain," the former controlling the archipelago via its Sunni allies in Iran while the latter administered the mainland—they nonetheless represent for today's Bahraini Shi'a population a common historical and ethical starting point that is both a symbol and a legitimizing force of their contemporary struggle for a greater influence over Bahraini society in the face of continued foreign domination. Thus they refer to themselves using the collective demonym *baḥārna* (sing. *baḥrānī*) in reference to their status as the "original" inhabitants of Bahrain, and in contrast to the Al Khalifa and their Sunni bedouin allies, who migrated from the Arabian hinterland and who only later, it is claimed, invented the modern designation "Bahraini" (that is, *baḥraynī*) as part of their effort to rewrite the country's Shi'i past.[6]

This seemingly semantic distinction would arise repeatedly during the process of conducting field interviews in Bahrain. Survey questionnaires, which employed the standard Arabic demonym "*baḥraynī*" throughout, regularly returned with lines drawn through the term and "*baḥrānī*" scribbled in the margins. Field interviewers, in particular those operating in the rural villages exclusive to Shi'a, recounted how they were lectured by respondents about Bahraini history and how the latter, when asked the survey question, "How proud are you to be a Bahraini?" replied matter-of-factly, "I am not proud of being *baḥraynī*; but I am very proud to be *baḥrānī!*"

The influence of this nativist discourse is made all the more powerful by the active effort by authorities to suppress it. Prominent books on the pre-Al Khalifa history of Bahrain, the royal family itself, and pre-independence politics are banned and subject to confiscation. This includes, for example, Fuad Khuri's afore-mentioned *Tribe and State in Bahrain*, which chronicles the nation's post-oil political transformation away from what he calls its prior "feudal estate system." In painstaking detail, Khuri describes how for more than a century prior to sweeping administrative reforms imposed by the British beginning in the 1920s, the island functioned as a complex of independent agricultural fiefdoms worked by structurally indebted Shiʻa and administered by absentee landlords from among the Al Khalifa and their Sunni tribal allies, the latter having been gifted considerable lands for their part in the Bahrain conquest. Despite being unsympathetic to Bahraini Shiʻa politically and ideologically, the book nonetheless is in high demand for its vivid account of nineteenth-century and early to mid-twentieth-century Bahrain—so much so, in fact, that I was requested upon my return to the United States to send a Bahraini friend but one gift: a copy of Khuri's book to replace one that had been confiscated from his father some time ago at Bahrain Airport.

Another significant account of the twentieth-century political history of Bahrain is found in the journals of Charles Belgrave, the British colonial officer who served in the position of personal "advisor" to the Bahraini ruler for some thirty-one years between 1926 and 1957, and who eventually came to be known simply as "*al-mustashār*—"the advisor." In the wake of British intervention just three years prior to replace recalcitrant ruler Sh. ʻIsa bin ʻAli with his son Hamad, Belgrave's extended appointment was meant to provide Bahrain with a measure of political continuity while he completed the task of modernizing the country's outdated bureaucracy, an initiative that effectively spelled the end of the prevailing feudal estate system and one therefore strongly endorsed by Bahraini Shiʻa but resisted by the ruling and wealthy elite. Belgrave's diary, then, consists of detailed daily reports of meetings and conversations with the ruler and various state officials, observations on Bahraini society, and other quotidian affairs.

Like Khuri's, this work too is banned in Bahrain. Or, more precisely, while selected excerpts from the diary were published in 1960 and again in 1972,[7] the original papers are said to reside in the Al Khalifa royal library and in any case have not been made available. However, in June 2009 unauthorized copies of the diary made their way onto the Internet, and were thereafter translated, made to be published, and imported for distribution by the Bahrain Centre for Human Rights (BCHR), itself an outlawed group of activists associated with Bahrain's minority Shirazi Shiʻa community. The printed copies were confiscated by the Ministry of Information, which informed the publisher of the government's decision to ban the book, barring any further imports.[8] But the damage was already done. The leaked version has persisted in the form of a massive, 2,302-page electronic document that has become required reading for the country's political activists.[9]

Yet, for all the controversy surrounding the Belgrave diary, and despite its constituting perhaps an embarrassing intrusion into the private lives and court politics of the ruling family, there is nothing in it that could be considered a direct attack on the Al Khalifa or that substantiates any heretofore unknown wrongdoing that must be covered up at all costs. No, the papers of Belgrave, like *Tribe and State in Bahrain*, are banned not for their contents per se, but for what they represent: namely information, and more importantly, contradiction of the "official," sanitized version of Bahraini history that the regime has worked hard to construct. Curiously, the same minister of information who enacted the Belgrave ban has herself published at least two separate editions of translated excerpts from the journals,[10] a coincidence that prompted the BCHR to speculate about the completeness and accuracy of the latter volumes and ask whether this might not help explain the ban on its own, presumably less-abridged Arabic translation.

Such promotion of an idealized Bahraini history goes beyond mere suppression of conflicting accounts, however, and pervades nearly all aspects of state-sponsored media and cultural displays. Linguist Clive Holes, for instance, has demonstrated that characters in the serials produced by Bahraini national television speak a distinct Sunni Arab dialect and ignore almost entirely the vernaculars of both the Baharna and the 'Ajam.[11] In similar fashion, the Bahrain National Museum, in its sprawling dioramas depicting pre-oil industry, gives a prominent place to the Sunni-dominated activity of pearl fishing while neglecting the quintessentially Shi'i agricultural sector, most obviously the widespread cultivation of date palms, which was the basis of the Baharna's existence for the centuries preceding the discovery of oil. Even Bahrain's prehistoric stone burial mounds, which some prominent Salafi politicians have suggested should be destroyed for their pre-Islamic origins, are represented quite extensively, with an entire full-sized mound and pieces of others having been reconstructed inside a large exhibit. The National Museum, naturally, is accessible via an exit along Al-Fatih Highway.

The nativism employed by Bahrain's Shi'a that has given rise to the present categories Bahrānī and Bahrayni—to the land's "original" Shi'i inhabitants versus their "foreign" Sunni oppressors—is therefore fueled in no small measure by the countervailing effort on the part of the state to downplay the nation's Shi'i past and, more generally, to obscure the details of Bahrain's pre-oil history while emphasizing its subsequent social and economic modernization and development under Al Khalifa leadership. That the authorities would expend such resources in rebranding the state away from its Shi'i roots, the Baharna reason, only goes to prove the validity of their version of the country's contested history and, by extension, the legitimacy of their attendant claims to a collective right in political decision making.

Today, however, there is being written a final, more ominous chapter to this Shiʻa counternarrative. It tells how the ruling family and its Sunni allies, having failed in their attempt to suppress and distort the true history of the country and so extinguish the embers of Shiʻis' legitimate political aspirations, have settled now on a more radical solution: the physical elimination of the Baharna's longstanding demographic majority through an organized program of so-called "political naturalization" of Arab and non-Arab Sunnis. Known popularly as "*al-tajnīs,*" this granting of Bahraini citizenship on a sectarian basis for political purposes is *primus inter pares* among the island's myriad points of contention, and it presents an instructive lesson in miniature on the group bases of political action in Bahrain.

Passports for Allegiance: Political and Demographic Engineering in Bahrain

Widespread accusations that Bahrain was attempting to alter its demographic balance through selective naturalization first surfaced in the run-up to the 2002 parliamentary elections, which were to be the country's first since the legislature was suspended in 1975 by the late emir, Sh. ʻIsa bin Salman Al Khalifa. Upon his death in 1999, his son Hamad ascended to the monarchy promising a general *rapprochement* aimed at easing political tensions and, more specifically, at ending a tumultuous decade of Shiʻa-state conflict punctuated by a wholesale Shiʻa uprising spanning 1994–1999.[12] The reestablishment of the parliament, then, was but one facet of an auspicious but ultimately illusory plan for political change outlined in a new National Action Charter, a written reform framework approved by popular referendum in February 2001. The document also mandated the release of political detainees arrested during the uprising, a general amnesty for political exiles, and the amendment of a repressive State Security Law to disband Bahrain's notorious State Security Courts.[13]

Prior to the vote, finally, Shaikh Hamad made a dramatic personal visit to the home of prominent Shiʻi religious leader Sayyid ʻAbdallah al-Ghurayfi, who received him along with Bahrain's most senior cleric and the spiritual force behind the 1990s intifada, Sh. ʻAbd al-Amir al-Jamri. There Bahrain's new leader signed and ostensibly agreed to a list of political demands stipulating that the lawmaking power of the new regime should reside in a democratically elected lower house of parliament, with any appointed upper chamber limited to a strictly advisory role. The widely circulated document, complete with photos capturing the act of signature, was seen as a coup for the opposition.[14]

The Charter was approved overwhelmingly: a full 89 percent of eligible voters were recorded to have taken part, with 98.4 percent in favor. It was approved so overwhelmingly, in fact, that Shaikh Hamad took its passage for a mandate to fashion a permanent constitution, indeed an entire new country, unilaterally. The non-negotiated document was promulgated on the first anniversary of the referendum in a flurry of royal decrees. The opposition was floored. Not only did Hamad renege on his public declaration by subordinating the elected Council of Representatives (*majlis al-nuwwāb*) to a royally appointed Shura Council with an equal number of seats and tiebreaking vote, but he also transformed the very state itself, declaring himself monarch of a new Kingdom of Bahrain. The Constitution of 2002 explicitly proscribed any change of this system, affirming that "it is not permissible under any circumstances to propose the amendment of the constitutional monarchy and the principle of inherited rule in Bahrain, as well as the bi-cameral system."[15]

Additional decrees promulgated later in the year further infuriated opponents. Decree No. 56 extended a previous amnesty order to the employees of the Ministry of Interior in the face of widespread claims of torture and other human rights abuses by those imprisoned during the 1990s, denying thousands of victims the opportunity of legal redress. Another, released in July, forbade the *majlis al-nuwwāb* from "deliberating on any matter or measure taken by the government prior to 14 December, 2002"—that is, prior to the inauguration of the National Assembly.[16] Yet another rearranged the country into twelve municipalities within five governorates, producing forty gerrymandered electoral districts ranging from 500 to 17,000 registered voters.[17] In the Sunni-dominated Southern Governorate, six members of parliament would represent some 16,000 voters, while a single district in the Shiʻi suburb of Jidd Hafs alone exceeded that number. Indeed, the entire Northern Governorate, a Shiʻa-populated region then home to 79,000 registered voters, was allotted a mere nine seats in parliament.[18]

Now just months away, the impending October parliamentary elections became a referendum on the new constitution. Municipal elections held earlier in May had seen a meager 51 percent turnout, a stark contrast to the near-universal participation of the prior year. Sensing its constituency's deep frustration with the government's now-unmasked "reform" agenda, the united Shiʻa bloc al-Wifaq opted at length to boycott the parliamentary vote despite an extremely successful showing in preceding local elections. The move temporarily averted an intra-Shiʻa schism that would occur four years later when al-Wifaq leaders would make the opposite decision to participate. Three other notable opposition societies—two secular groups and one affiliated with the Shirazi Shiʻi *marjaʻ* Ayatollah Hadi al-Mudarrisi—followed suit, and so voter turnout reached just 53 percent in the first round and a dismal 43 percent in the second.[19] The resulting parliament would consist wholly of Sunni Islamic societies and pro-government "independents" from historically aligned tribes and families.

It was under this charged political backdrop that rumors resurfaced about a concerted government effort to alter Bahrain's demography. Similar claims had been made as early as May 1998, when in the darkest days of the Shiʻa intifada the *Financial Times* reported:

Critics of the government say one sinister development is the building by the ruling family of a *cordon sanitaire* around itself by giving nationality to between 8,000 and 10,000 Sunni families from Jordan, Syria, Pakistan and Yemen, whose men, working in the security services, would be loyal to the al-Khalifa family should unrest break out again on a scale which can no longer be contained.[20]

This time, however, the accusers would offer hard evidence in the form of a 17-minute video interview allegedly conducted in June 2002 with members of the al-Dawasir tribe of Saudi Arabia's Eastern Province, who tell how they were solicited to obtain Bahraini nationality and public housing in the run-up to the elections. Not only were they granted passports in just a matter of months, said the al-Dawasir tribesmen, but for Bahrain's 2002 elections they were gathered and driven to a polling station on the King Fahd Causeway linking Bahrain to Saudi Arabia in order to cast their votes. Among the al-Dawasir, they estimated, perhaps 20,000 had received this dual citizenship, which they said was now being extended to other tribes around Dammam. Those interviewed were dually able to produce Bahraini passports, national identification cards, and addresses in the country.[21]

When the documentary was aired publicly in July 2003 at a meeting of opposition groups in Manama, the reaction was not primarily one of surprise but of vindication. Observers had long argued that a June 2002 royal decree allowing other GCC citizens to hold dual-Bahraini citizenship and vice versa would be put precisely to this end.[22] All the same, the release of the film and subsequent public outcry did prompt the formation of a parliamentary committee charged with investigating the scandal, though in a clear act of sabotage its members were forbidden from examining citizenship cases prior to the December 2002 establishment of parliament as well as those "special" cases falling under an exception granted the head of state per Bahrain's 1963 Naturalization Law.[23] Since all of the cases in question could be made to fall under one or the other category, the matter was effectively closed for official discussion.

It was closed, that is, until the sudden appearance three years later of a now-notorious leaked exposé by a British national of Sudanese origin, Salah al-Bandar, working then as an advisor to the Cabinet Affairs Ministry.[24] His 216-page document claimed to outline a clandestine network of activists and officials working to undermine the overall political position of Bahraini Shi'a. Implicated as leader of the campaign was Shaikh Ahmad bin 'Atiyatallah Al Khalifa: then minister of cabinet affairs, then and currently minister for follow-up at the Royal Court, and founder of the High Committee for Elections.[25] The scandalous report, discussion of which was quickly barred in the media and parliament, described a wide-ranging effort based upon a new understanding of Bahraini Shi'a as not merely a *political* problem but a problem foremost of national security.

According to al-Bandar's report, this new conceptualization—as well as the anti-Shi'a program to which it would give rise—reflected the recommendations of a 2005 study written by an Iraqi academic under commission from the Bahraini government titled "A Proposal to Promote the General Situation of the Sunni Sect in Bahrain."[26] The text, which al-Bandar appended to his dossier in a documentation section, blames "the rise of sectarian conflict" in Bahrain—a conflict "between the Sunni sect on the one hand and the Shi'i sect on the other"—on "the existence of an unspoken agenda on the part of Shi'a movements to control Bahraini society, and [these] ambitions may extend to taking over the reigns of power in the country." This situation, the essay continues, is the product of "the historic changes that threaten the Arab Gulf region [as a consequence of] the fall of the former Iraqi regime." Thus, it concludes, Bahrain's case is that of post-2003 Iraq:

> the marginalization of Sunnis and the lessening of their role in Bahrain is part of a larger regional problem, whereas [our] sons of the Sunni sect in Iraq face the same problem, meaning there is a direct correlation between [the Iraqi situation and] the marginalization of the Sunna in the Gulf countries, and their marginalization in Bahrain in particular. Thus there is a dangerous challenge facing Bahraini society in the increased role of the Shi'a [and] the retreat of the role of the Sunna in the Bahraini political system; namely, the problem concerns the country's [Bahrain's] *national security*, and the likelihood of political regime change in the long term by means of the present relationships between Bahrain's Shi'a and all the Shi'a in Iran, Iraq, Saudi Arabia's eastern region, and Kuwait.[27]

To combat this long-run existential threat posed by Bahraini Shi'a acting in concert with their co-sectarians across the Gulf region, the paper advocates that the state undertake a multifaceted program designed to dampen the group's influence in Bahrain and its ability to elicit sympathy and support from audiences abroad. As summarized in the proposal, the overall "goals of the project" would be threefold: to

1. Protect the gains achieved by the national reform project launched by His Majesty the King;
2. Protect the Kingdom of Bahrain from any external interference from regional powers in which the role of the Shi'a has increased, such as Iraq and Iran, and this in order to prevent the destabilization of the [Bahraini] political system; and
3. Protect the Sunni sect in the Kingdom of Bahrain from any Shi'a attempts at marginalizing [it] in the political system or within Bahraini society in general.[28]

In pursuit of these aims the document urged the Bahraini government to undertake a coordinated plan of, *inter alia*, increased naturalization of Sunnis, infiltration of Shi'a nongovernmental organizations, establishment of parallel civil society and human rights groups to counter the effective media campaigns of Shi'a activists inside and outside Bahrain, and even a Shi'i-to-Sunni religious conversion program.

It was this agenda, then, for which the network of Sunni politicians and officials led by Sh. Ahmad bin 'Atiyatallah was allegedly organized. As evidence of this central claim, al-Bandar's report documents bribes and payments totaling more than one million Bahraini dinars (nearly $2.7 million) dispersed among various members of an electronic group, a media group, an intelligence team, a newspaper, and other para-governmental organizations funded under the initiative. The documentation, comprising some 80 percent of the report, includes photocopies of hundreds of receipts, letters, bank statements, and account sheets, and outlines the personal relationships linking those involved. Bahrain's pivotal 2006 parliamentary elections, then only one month away, were consumed by talk of the "Bandargate" scandal.[29]

However, like the allegations three years earlier of improprieties in granting citizenship and of electoral fraud, the revelations unleashed with al-Bandar's bombshell could sustain only a temporary outburst of public protest due to a swift government containment effort. Indeed, of those alleged to have been involved in the plot, it seems that the only political casualty was al-Bandar himself, who upon the release of his report was promptly dismissed, arrested, and deported. Ahmad bin 'Atiyatallah, on the other hand, to the universal consternation of Bahrain's Shi'a, not only retained his position as cabinet affairs minister but in fact seemed only to grow in influence within the ruling family for his perceived skill in managing the Shi'a threat. In subsequent years he would be granted additional portfolios, culminating in an April 2011 appointment as minister for follow-up at the Royal Court, a position affording "wide-ranging powers."[30]

Whatever the case, public discussion of the matter would be cut short abruptly following a press gag order handed down less than two weeks after the story began to appear in newspapers, at a time when even the government-affiliated *Gulf Daily News* was forced to lead with the dramatic page-one headline: "BANDAR-GATE!"[31] Crowds of protesters marched in "anti-political naturalization" rallies through Bahrain's posh Seef shopping district, and one hundred prominent political figures, nearly all of them Shi'a, authored a public petition to King Hamad "appealing to [him] to give a public speech to the common citizens to answer all those dangerous queries and to announce what will be done in regards to that sectarian plan and secret organization that is implementing it."[32] But no official government comment, to say nothing of a formal address by King Hamad, would be forthcoming.

Debate surrounding al-Bandar's report, in particular the "dangerous query" as to the ultimate source of the funds made available to Shaikh Ahmad and his associates, would be confined to private diwans until the swearing in of the new parliament the next year. Members of al-Wifaq, which had reversed its previous electoral boycott to capture 17 of 40 seats, walked out of the chamber just four months into the session in May 2007 in protest when their resolution to question the cabinet affairs minister twice failed to muster the additional four votes needed to pass. Prompted by newly published government data indicating a dramatic jump in the number of Bahraini citizens—an increase, according to critics, that could not have occurred without mass naturalization[33]—al-Wifaq MPs tried again to quiz Ahmad bin 'Atiyatallah in March 2008. The action provoked a three-week stalemate that ended with the hasty replacement of a longtime parliamentary legal advisor with another who immediately ruled the entire motion unconstitutional. "Parliament's future is blurred," remarked one al-Wifaq parliamentarian, "a crippled and unworthy institution, which pokes its eyes with its own fingers."[34] Bandargate, and with it a decisive chapter in the ongoing controversy surrounding *al-tajnīs*, was over.

A Myth of Their Own: The Bahraini Shi'a as an Imagined Fifth Column

Whether one inclines to believe al-Bandar's account or dismisses it as an elaborate, calculated forgery, the viewpoint it embodies, the notion that Shi'a today represent a transnational political front to be managed and contained by Gulf Arab governments, is certainly no fiction. Even the seemingly inflammatory "Proposal to Promote the General Situation of the Sunni Sect in Bahrain" says little more than did, for example, Jordan's King 'Abdallah II in a now-famous interview with the *Washington Post* in December 2004. Therein he characterized the newly empowered Shi'a of Iraq as part of a menacing "Shiite crescent" that could extend all the way from Syria and Lebanon through Iraq and Iran and into the Arab Gulf. Such a front of Shi'a-led governments and movements would constitute a destabilizing force, King 'Abdallah warned, from which "[e]ven Saudi Arabia is not immune. . . . It would be a major problem. And then that would propel the possibility of a Shiite-Sunni conflict even more, as you're taking it out of the borders of Iraq."[35]

The news media and several prominent books, exemplified by Vali Nasr's 2006 work *The Shia Revival*, offered extended elaborations of King 'Abdallah's broad anxiety, weaving disparate events across the Islamic world into a coherent narrative of coordinated Shi'a emboldening that included a combative new Iranian president hell-bent on erecting a military nuclear program; an Iraqi state transforming into an Iranian puppet; a confident Hizballah in Lebanon and Hamas in Gaza, each prepared to take on the Israeli army using sophisticated hardware from Iran; and a set of Arab Gulf states looking increasingly vulnerable to Shi'a irredentism. As Louër explains, the power of this "Shiite crescent" concept "no doubt lay in its ability to sum up in a short formula the spontaneous perception of the Shias by the majority of the Sunnis: people united by a corporate solidarity beyond national borders and subservient to Iranian expansionism."[36]

In the case of Bahrain, this interpretation had prevailed since long before its unofficial coinage by the Jordanian king, to say nothing of the uprising of February 2011. It was during a May 2009 interview with an outspoken Salafi preacher and three-term parliamentarian that I would hear the clearest articulation of what is essentially the Sunni counterpoint to the Baharna's "myth of golden age" describing the premodern history of Bahrain. Shaikh Jasim al-Saʿidi, a well-known imam who delivers regular Friday sermons at a mosque near his home, agreed to meet me at his weekly public *majlis*[37] despite expressing some hesitation to our intermediary that as an American, and given his reputation for controversial remarks, I may attempt to misrepresent his words. Indeed, he was fresh off a showdown in which members of al-Wifaq nearly succeeded in stripping him of parliamentary immunity in preparation for prosecution in response to a sermon in which he reportedly "compared some Shiites of Bahrain, without naming their sect, to 'the sons of Zion bent on acts of destruction and sabotage.'"[38] It was for comments such as these that Sh. al-Saʿidi was deemed "too extreme" to stand for elections even with the main Salafi society al-Asalah, itself not known for its liberality. So, in each of Bahrain's first three parliamentary elections—in 2002, in 2006, and again in 2010—he ran and won as an independent. As it happened, however, no manipulation of the shaikh's words would be necessary.

The gathering, a popular forum attended that evening by some 50 to 75 guests, was organized on this occasion around a particular piece of legislation—the contentious Sunni Family Law that codified important religious regulations in civil law[39]—agreed earlier that day after much argument, delay, and, most importantly, government pressure for its passage. Sh. al-Saʿidi, known to be a staunch regime loyalist, offered a spirited sermon in defense of the measure on religious grounds, arguing that civilian codification of the Islamic Law is implied by the very writing of the Qur'an, prophetic Sunna, and Hadith. Any notion to the contrary, he railed, is an obvious "influence of secularism, whose forces continue to wage war against Islam." He then fielded several questions from the audience, to which the ultimate answer was that in such matters people ought to follow their religious leaders; and that if the latter are wrong, they will be held to account before God.

At length Sh. al-Saʿidi handed the microphone to another imam seated next to him to deliver some concluding "news items." The first involved the reading of a report by a Bahraini scholar revealing the "true" historical populations of Sunna and Shiʿa in the country disaggregated by region, statistics that demonstrated to what extent the Shiʿa are wrong in asserting that they have always been a majority; and that it was *they*, not the Sunnis, who have achieved numerical superiority as a result of naturalization. (Incidentally, Sh. al-Saʿidi is known by many Shiʿa detractors as "*shaykh al-mujannasīn*," "shaikh of the naturalized," for his reputed ideological support for *al-tajnīs*.) The audience was made to show shock and consternation at these data as the speaker read aloud the former populations of Sunna and Shiʻa across various parts of Bahrain. Then, by way of closing, and one assumes not coincidentally, the speaker noted that the following day there was planned an anti-naturalization meeting to be held in the Sunni neighborhood of ʻArad. It was organized by Ebrahim Sharif of socialist-leaning Waʻad—a Sunni no less—and members of al-Wifaq. God willing, he said, it would be cancelled (that is, by the Interior Ministry), but, just in case it was not, those present should plan to protest the site and also block access to it with their cars.

After the *majlis* I thanked Sh. al-Saʻidi for the opportunity to speak with him and said that I hoped one of my questions, his take on a recent royal pardon of 178 Shiʻa detainees arrested during antigovernment rioting over previous months, had not been too sensitive. At this he launched a discussion about those who had been imprisoned and then released, calling them "terrorists" doing "the work of Iran." Bahraini Shiʻa, he explained, illegally smuggle their co-sectarians across the Gulf from Iran in large boats. Once they arrive, they are taken to *mawātim*[40]— holy places that the police cannot enter—until they have been able to learn sufficient Arabic to apply for Bahraini citizenship under the pretext that they had been residing in the country for decades but had never naturalized.[41] "A very dangerous situation," I agreed as I left.

Here, then, we find the basic outline of the Sunni rejoinder to the Shiʻi political history of Bahrain: in the first place, historically speaking the Baharna have never formed a majority of the island's population, having come close to doing so only recently and through decades of immigration from the Iranian mainland, al-Qatif and al-Hasa, Iraq, and Kuwait. Thus their complaints of political underrepresentation and discrimination, made on majoritarian and nativist grounds, are ill-founded and disingenuous. Moreover, even if Shiʻa did outnumber Sunnis today in Bahrain, why should they expect an equal or greater share of state benefits while they show themselves to be lesser citizens? Their highest religious leaders, who by their own doctrine also wield supreme political authority, are not Bahrainis or even Arabs but Persians living in Iran and Iraq, and yet these individuals still see fit to interfere in the internal affairs not only of Bahrain, but also those of Iraq, Lebanon, Yemen, Kuwait, Syria, and countries throughout the region. So, if the Bahraini government is afraid to let the Shiʻa serve in the military or sit on the royal court, who can blame them? They have already attempted a coup once, with Iranian help,[42] and continue to burn tires, kill police officers, and in general cause trouble for society every day. And all this despite a generous if measured program of reform initiated by the king in 2001 that has improved the political situation dramatically: Bahrain now has elections, a parliament, many Shiʻa ministers—what more do they want?

One might suppose that such a combative response to the Shiʿa "myth of golden age," one that echoes precisely the suspicion of the "Proposal to Promote the General Situation of the Sunni Sect in Bahrain" and the idea of a united "Shiʿa crescent," would be confined to the personal views of noted "extremists" like Sh. al-Saʿidi. Yet far from being an outlying opinion, this perception of the Shiʿa—at a minimum, of Shiʿa political dissidents—as constituting a veritable fifth column in Bahraini society I found to be common among ordinary citizens and the Sunni political elite alike, well before the events of early 2011. Dr. ʿAli Ahmad, Secretary General of the Muslim Brotherhood-affiliated parliamentary bloc al-Manbar al-Islami (hereafter al-Manbar), spoke of a similar link between the foreign ambitions of Iran and its assumed agents in Bahrain, saying,

> As for the polarization in the country, this is the result of Iran's using some people in Bahrain for its own interests—that is, to achieve Iranian control of Bahrain. Before the Iranian Revolution there were no sectarian problems in the country. I attended school with and lived next to Bahraini Shiʿa and didn't even know it because [the situation] wasn't politicized. If democracy in Bahrain becomes divided along sectarian lines such as in Lebanon or even in Iraq, then the situation will be bad.[43]

Here the explicit words of Sh. al-Saʿidi are replaced by the more diplomatic "some people in Bahrain," a standard euphemism whose meaning is betrayed by the subsequent reference to "Bahraini Shiʿa." The implication in any case is clear: there are Shiʿa inside Bahrain working at the behest of the Iranian regime and in pursuit of the same agenda that has driven it for the past three decades, namely the exportation of the Islamic Revolution to the Arab Gulf countries in general and to Bahrain in particular.

A less restrained version of this argument was articulated by another Salafi legislator, ʿIsa Abu al-Fath, who when asked to name the biggest challenge currently facing the country concluded by saying,

> The other big challenge, facing not only Bahrain but the entire Gulf, is Iran, which wants to recreate the Persian Empire throughout all of the Gulf areas from Kuwait to the UAE. It will be a nightmare for everyone as Iran continues to grow in power, and the U.S. will be too afraid to do anything about it.[44]

Yet despite these Iranian pretensions, I noted, the Gulf seems still to enjoy more security and stability than other Arab countries. How can this be? True, he admitted,

> However, the terrorist attacks that do occur are mostly done by fighters trained in Iran. How do most of the Taliban fighters and Arabs in Afghanistan get there? Just pay 2,000 dollars and they will get you a ticket to Tehran and from there you can go to Afghanistan. Even the 9/11 attacks—those Saudi hijackers were trained in Iran. . . .

Even the Bahraini Shiʻa train in Iran or with Iranian help in Lebanon with Hizballah. During the 2006 war between Israel and Hizballah, the Bahraini Interior Ministry went to Lebanon to try to evacuate all the Bahraini citizens stuck there, and they found a lot of Bahrainis fighting with Hizballah. The Bahraini Ministry of Immigration found that in just one year over 20,000 Bahrainis traveled to Syria for "vacation." Hizballah is also buying any land it can get in the GCC countries with money from Iran; and Iran's money in turn comes from the *khums*[45] [one-fifth] tax that Shiʻa pay to the mullahs in Iran. We in al-Asalah tried to introduce a 2.5% *zakat*[46] tax for Bahrain in parliament, and the Shiʻa [i.e., al-Wifaq] opposed it; yet they pay 20% of their incomes to Iran. And then they complain of being poor. If they are so poor, how can they afford to pay?

Rhetoric of this kind, which has in common the belief that some or all of Bahrain's Shiʻa are knowing pawns in a larger game of Iranian geopolitics, blindly following whatever orders arrive from Najaf, Qom, or Tehran, even escalates to ascribe to them a more menacing role: that of principal graduated from mere agent. It is one thing to contend that Bahrain's Shiʻa community is exploited as an instrument of domestic subversion by a scheming regime in Iran, serving the latter's political agenda as its local representative. It is more serious, however, to suggest that it constitutes in its own right an independent center of activism projecting political destabilization elsewhere; that the Shiʻa of Bahrain are themselves a sort of Iran vis-à-vis the rest of the Arabian Peninsula—no longer students but the teacher.

Yet this is precisely the accusation that surfaced in August 2009 in a controversial interview with the aforementioned Sh. al-Saʻidi printed in the Saudi daily *Al-Sharq Al-Awsat* (*The Middle East*). In it, he claimed to have learned that members of al-Wifaq had met secretly inside Bahrain with a high-level representative of Yemen's Huthi movement, a separatist faction fighting to restore the country's Zaydi Shiʻa imamate. (By some accounts, the group has more recently adopted Twelver Shiʻism, a fact taken as evidence of its strengthening ties to the Islamic Republic.) The Huthis' ongoing conflict with the central government,[47] portrayed in the Arab and Western media as a "Shiʻa insurgency" supported ideologically and materially by Iran, had just resumed for a sixth iteration. Under this backdrop Sh. al-Saʻidi revealed,

> We have confirmed information that members of the al-Wifaq bloc met in Bahrain with key political figures with strong ties to the Yemeni Huthis, and this but a few months prior to the outbreak of the [sixth Saʻadah] war between the Yemeni government and the Huthis, which raises a lot of questions about the connection linking the Huthi insurgents in Yemen and the Wifaqis in Bahrain.[48]

The meeting, he continued, took place

> with the knowledge that this Huthi [representative] has a [criminal] past and a suspicious history in the Yemeni Republic, where previously he was arrested

on the back of visits and conferences he participated in [inside] the Iranian Republic, which embraces the errant Huthi [i.e., Twelver Shiʻi] ideology and funds it through an octopus-like network of cells distributed throughout all of the Gulf and Arab countries.[49]

Al-Saʻidi's insinuations are noteworthy on several accounts. First, they reinforce the idea of a transnational Shiʻa front united in religious solidarity. Why, otherwise, would the Shiʻa of Bahrain, represented here by al-Wifaq, have any connection to a few thousand individuals living in an isolated mountain region of northern Yemen? Indeed, if anyone should be suspected of taking an interest in their cause, a more natural choice would be the sizeable Ismaʻili Shiʻa community concentrated but a few miles away across the Saudi border in Najran.[50] That Bahrainis would sympathize with their Shiʻa brethren in Yemen, one is left to infer, is a foregone conclusion owing to their inviolate bond as co-sectarians. This view is further implied by some pregnant wordplay that draws a clear parallel in Arabic between the "Huthis in Yemen and the Wifaqis in Bahrain," where the adjective "insurgents," almost always used in the media to describe the Huthis, can be interpreted as applying to both them and al-wifāqiyyīn, a nonstandard eponym used here as an epithet suggestive of an ideological cause.

The other, more remarkable aspect of these claims by al-Saʻidi is that, rather than accuse al-Wifaq of meeting with the Huthis as an intermediary of Iran, he credits the group with operating its own agenda outside Bahrain, interfering in the affairs of its neighbors and fomenting Shiʻa irredentism as a veritable Iran-in-miniature on the Arabian mainland. He even goes so far as to imply that al-Wifaq had some role in the resumption of hostilities in Yemen, seeing as how their meeting with Huthi representatives took place "but a few months prior to the outbreak of the war," a coincidence, Sh. al-Saʻidi hints coyly, that "raises a lot of questions about the connection linking the Huthi insurgents in Yemen and the Wifaqis in Bahrain." Such allusions to direct assistance—and military assistance at that—in aid of Shiʻa factions abroad are almost always reserved for Iran proper, whom the former Yemeni government of Ali Abdallah Saleh routinely accused of fomenting the conflict in Saʻada. That the same role should be ascribed now to al-Wifaq, the most moderate if largest of Bahrain's Shiʻa political formations, betrays a grave apprehension on the part of Bahraini and Gulf Sunnis, including Sunni royal families, for whom the danger is not al-Wifaq qua political bloc but al-Wifaq as a symbol of the increasing Shiʻization of the Arab Gulf.

Yet even al-Saʻidi was loath to spell all this out explicitly, at least not publicly, in the manner of the Jordanian king and his evocative "Shiʻa crescent." For this, however, one may turn to Yemen's former president of 33 years, Ali Abdullah Saleh, himself a Zaydi Shiʻi and also not one to mince words. In a primetime interview with the Saudi news channel Al-Arabiya in March 2010, at a time when the sixth war for Saʻadah threatened to spiral into a full-scale regional conflict after the Huthi rebels crossed into Saudi territory, Saleh was asked about his knowledge of foreign support for the group. In the first place, he began, the attempt to drag Saudi Arabia into the war is proof in itself of outside involvement, since the Huthis would not have taken such a step on the basis of military considerations alone. Said Saleh,

> I am certain that more than 80 to 90 percent of it is foreign encouragement, in order for countries of the region to settle their scores with Saudi Arabia, to preoccupy Saudi Arabia, and to send a message to Saudi Arabia via these Huthi elements. [I say this] because we don't have a problem with [rebel leader 'Abd al-Malik] al-Huthi—al-Huthi . . . what problem is he? [But] al-Huthi now has a foreign ideology: let's say, [one] based on Twelverism, [while] he is Zaydi. We in Yemen are Zaydis and Shafi'is. We have no problem [between us]. The entry of the Twlever sect, introduced to the Huthis [from outside], is something new . . . something new. We aren't against the Twlever sect . . . the Shi'i [sect] anywhere. We aren't against [them]. We believe in a diversity of sects. But we reject its being imposed on our country, or [that] we [should] adopt it. Because for thousands of years in Yemen we've been Shafi'is and Zaydis; and there is no dispute between Shafi'is and Zaydis. And this new, errant sect will pay . . . say, will pay the price [for promoting sectarian strife].[51]

When asked directly to name the "countries of the region" known to be giving the Huthis such "foreign encouragement," Saleh hesitated but went on to acknowledge that some of these "foreign elements" can be found in "Saudi Arabia, London, and America." He explained,

> They are countries and individuals . . . individuals in countries . . . in most of the countries in the region. They are all those who sympathize with the Huthis in the name of Twelverism . . . in the name of Shi'ism. So any Shi'a in the region, they are the ones that sympathize with and raise some funds to support the Huthis.[52]

Here, then, are the words that Sh. al-Sa'idi in his reproach of al-Wifaq intended but could not say: the "Wifaqis" have forged a relationship with the Huthis of Yemen not by chance, not because they are by nature a meddlesome group that tends to interfere in other countries' affairs, but because they identify and commiserate with them as fellow Shi'a, as a people who itself complains of political repression born of religious discrimination. In this respect Saleh's seemingly out of place reference to Huthi supporters in "London and America" is instructive, as the geographical remoteness of both locations contrasts markedly with their importance as new global centers of Shi'a activism,[53] giving the impression that wherever one finds a Shi'i, whether in Dammam or Detroit, there he finds a friend of all other Shi'a, a loyal soldier ant who when he senses any of his brood in trouble runs instinctively to their defense. This "new, errant sect" that has infiltrated Yemen, upsetting "thousands of years" of religious harmony, has arrived therefore not from Iran only, the most obvious party looking to "settle [its] score" with Wahhabi rival Saudi Arabia, but through the help and support of Shiʿa everywhere, where "Shiʿa" is understood to refer specifically to Twelver Shiʿis.[54] That members of al-Wifaq might be involved, then, is just the tip of the iceberg.

Whether or to what extent such recriminations reflect reality—whether the claims of Sh. al-Saʿidi, Yemen's president, Jordan's king, or the Bahraini parliamentarians quoted above—is not of real concern. The decisive point is the degree of apprehension itself, this palpable unease among Sunni leaders and citizens alike at what is perceived as the rebirth of Shiʿa oppositions across the Arabian Peninsula in a seemingly coordinated political mobilization that harkens back to the early days of the Islamic Revolution. Accordingly, as Louër persuasively argues, the supposed "Shiʿa revival" is as much an artifact of changing threat perceptions as it is a result of Shiʿa initiative.[55] She writes,

> [I]t is through the representation that it aroused in the Sunni psyche and not through the modification of the Shia agenda that the regional context played a role in moving the Sunni/Shia relation. For the Shias, the new context only adds to the tools at their disposal to continue with their previous strategy. It is the Sunnis who now feel under siege.[56]

Hence, at a time when the entire region has at least one eye fixated on Iran, the undisputed if inadvertent winner of the U.S.'s "New Middle East" project, most Gulf ruling families recognize a new domestic menace in those seen to be divided by competing national and religious-cum-political loyalties. If the 1990s was the decade for most Gulf monarchies to combat Sunni extremism, the post-Iraq War era has been one of managing the Shiʿa, including Shiʿa frustrations, as a way of checking Iranian influence.

Yet, whereas this reprioritization has recently led to some positive developments in the two Gulf states other than Bahrain with substantial Shiʿa populations—a policy of what Louër calls "relative religious recognition"[57] in Saudi Arabia and Kuwait—Bahrain has found itself unable to arrive at a similar compromise with its Shiʿa citizens. In the first place, the Shiʿa of Bahrain face less religious discrimination at the institutional level than those of Saudi Arabia or Kuwait, there being, for example, already separate Maliki (Sunni) and Jaʿafari (Shiʿi) sections of the governing *sharīʿa*, Islamic courts, and religious endowments.[58] More importantly, unlike in Saudi Arabia and Kuwait, the Shiʿa of Bahrain do not represent an irksome minority but a full demographic majority, and one for whom, as Louër remarks, "a politics of merely religious recognition cannot substitute for a genuine democratization policy."[59] In the case of the Baharna, then, the demand is not for *religious tolerance* as practitioners of a particular faith but for *political equality* as members of a societal group whose members seek substantive influence closer in line with their demographic status; and whose very identity and institutions serve as important focal points for mass coordination in pursuit of this agenda.

Bahrain's Catch-22

One easily perceives, then, the extent of the dilemma facing Bahrain's rulers, who face a situation in which the very act of political concessions, meant to pacify Shi'a opponents and preclude their turning to Iran and other external actors for support, would itself be regarded as opening the door to outside influence. For, from the standpoint of the Al Khalifa, the intermediate goal of quieting the Shi'a—which could be bought only by agreeing to their demands for major constitutional reform, an end to political naturalization, and equal opportunity in public sector employment, including in the military and power ministries—is necessarily at odds with the primary objective, indeed the motivation for the entire exercise, which is to ensure the security of the regime. Given the choice between a Shi'a population that is politically agitated but militarily impotent and one that is politically satisfied but strategically better positioned within the government apparatus, Bahrain long ago decided upon the former. For, while some Shi'is may be driven closer to Iran, still they could never, even with Iranian help, pose an existential threat to Al Khalifa rule of Bahrain—particularly so long as the U.S. Fifth Fleet remains based at the island. In Bahrain, it turns out, regime stability entails not political tranquility but its opposite.

As we return then to Bahrain *qua* rentier state in light of Bahrain *qua* divided state, it is clear in what ways the latter must revise our understanding of the former. Mutual suspicion—the feeling among Shi'a of political and possibly (if those behind *al-tajnīs* had their way) physical disenfranchisement at the hands of foreign Sunni occupiers and their co-sectarian supporters, and the perception among Sunnis that the so-called Baharna are more akin to Iranians than to loyal Bahraini citizens—such mutual antagonism demonstrates how a class-based politics can indeed emerge in rentier societies, supplanting individual jockeying for royal patronage as the dominant political *modus operandi*.

Rather than compete independently for a greater personal share of state benefits, in countries with a significant ascriptive group cleavage, such as exists in Bahrain, Saudi Arabia, and Kuwait, citizens tend naturally to coalesce into broad factions competing over the very state itself. The key battles of politics are fought not along distributive lines but along the very defining lines of the regime: the nation's history and cultural identity, the bases of citizenship, and the conditions for government and martial service. To be sure, in a society where it is a matter of significant debate whether the true citizen is a "Bah-RAY-ni" or a "Bah-RAH-ni," it is clear that the rentier politics of allocation has taken a back seat to a group struggle over status and national ownership; that, as Horowitz says of ethnically divided societies, "the symbolic sector of politics looms large."[60]

It is equally apparent, under such circumstances, why the traditional pressure-releasing levers of the rentier state here lack the effectiveness they might otherwise have. In Bahrain sectarian division serves to handicap the government by at best making these options less efficient, at worst by taking them off the table altogether. "Every citizen" of a rentier state, Beblawi assures us, "has a legitimate aspiration to be a government employee; in most cases this aspiration is fulfilled." Though his qualification "most cases" is ambiguous, it is certain that one instance in which this aspiration will *not* be fulfilled is when a government harbors suspicions of disloyalty with regard to a prospective employee. And what if these suspicions extend to a full majority of a country's indigenous population? Then the state must fill shortfalls in the ranks of the police, the military, and sovereign ministries with individuals whom it does trust, often "non-partisan" foreigners imported specifically for the purpose. In short, this state begins to resemble Bahrain and other Arab Gulf regimes: employment itself being a political tool, those whose political allegiance is doubted are systematically excluded from large swathes of the public sector; and for every one individual undeserving of service, governments reason, a dozen can be recruited from Yemen, Syria, or Baluchistan.[61]

In moderation this situation may pose few problems for regimes, begetting nothing more than a small percentage of the population who must look to private industry for work or who perhaps remain underemployed and individually disaffected. But extend it to half of all citizens, indeed the very half that would tend toward government opposition even in the best of economic conditions, and one quickly runs the risk of systematic dissent by members of the excluded out-group. In a survey published in September 2003 of 32 ministries and the state-run University of Bahrain, the Bahrain Centre for Human Rights found that

> of 572 high-ranking public posts . . . Shiite citizens hold 101 jobs only, representing 18 per cent of the total. When the research was conducted, there were 47 individuals with the rank of minister and undersecretary. Of these, there were ten Shiites, comprising 21 per cent of the total. These do not include the critical ministries of Interior, Foreign [Affairs], Defense, Security, and Justice.[62]

More recently, the same BCHR revealed in a March 2009 report that according to a confidential list of over 1,000 employee names obtained from Bahrain's National Security Apparatus and circulated online, a mere 4 percent were Shiʻa, while 64 percent were "non-citizens, most of Asian nationalities."[63] My own nationally representative survey fielded in early 2009 obtained a similar result: not a single Shiʻi of those randomly sampled for interview reported working for the police or armed forces. Compare this to 13 percent of the 131 total working Sunni households that gave occupational data. In sum, even a cursory look at patterns of public-sector employment in Bahrain is enough to show that, at least in this rentier state, one must revise the familiar line that "every citizen has a legitimate aspiration to be a government employee" by qualifying the nature of both employment and employee.

Of course, one need not rely in these conclusions on the likes of anonymous Internet reports supplied by the opposition. For one can readily glean as much from public officials themselves, who, while they deny any specific cases of sectarian-based discrimination, seem in their comments to agree with the general sentiment. In an interview with the *New York Times* in March 2009, the chairman of Bahrain's parliamentary Committee on Foreign Affairs, Defense, and National Security, 'Adal al-Ma'awdah of al-Asalah, replied when asked about Shi'a claims of exclusion from the armed forces, "There are so many riots, burnings, killings, and not even one case is condemned by the Shiites. Burning a car with people inside is not condemned.[64] How can we trust such people?"[65] My own contacts echoed this reasoning. Samy Qambar, a parliamentarian from al-Manbar, said in regard to the Shi'a complaints,

> [S]adly, the Shi'a feel that they are a majority of the population and therefore entitled to have a greater presence in the government and army and police, while the government feels these posts should be filled with people who they can trust and who are loyal to them, not with people from the opposition.

Even top government officials make the same acknowledgment. Hassan Fakhro, then and currently minister for industry and commerce, admitted during anti-naturalization protests following the release of the Bandar Report, "There is a lack of confidence between the ruled and the rulers. It is not unusual. There is a small percentage who do not have loyalty to the state. Sometimes, for good reasons, you have to be careful who you employ."[66]

And careful Bahrain and other Gulf states are. Applicants for "sensitive" positions within the police, military, and bureaucracy are required to include a "certificate of good history and conduct," a document issued by the police to verify that an individual has no prior record of arrest or detention, including for political reasons.[67] A difficult hurdle to overcome for one accustomed to near-daily street demonstrations for the past two decades, the requirement has the effect of discouraging if not precluding ordinary Shi'a applicants for all but a limited set of "non-sensitive," low- and intermediate-level positions within state institutions.

Yet, more significantly, Fakhro's words summarize well the basic problem of public-sector employment as seen from the standpoint of Bahrain's rulers, or indeed from that of any regime distrustful of a particular (and outwardly identifiable) subset of its population: absent a reliable way to distinguish a good prospective employee from one lacking "loyalty to the state"—for even a clean past is no guarantee save for that one is prudent—does one choose to exclude the class of "disloyals" at the greatest possible level of abstraction (say, on the basis of sectarian membership) but with the largest margin for error? Or, instead, attempt to fish them out individually with the knowledge that one or another may slip by? In its choice between a trawler and a butterfly net, the Bahraini government has settled decidedly upon the former instrument, casting a general web of suspicion upon all Shi'a as a class of citizen, and accepting the collateral damage of whatever "loyals" may be inadvertently caught up in the mesh. It is this collateral damage, this lost opportunity and foregone co-optation due to sectarian distrust, that is a central feature of the dysfunctional rentier state that is Bahrain.

The causal direction of the standard rentier formulation must accordingly be reversed: it is not public-sector employment that secures political allegiance; rather, it is political allegiance that tends to secure public-sector employment, especially when the work in question carries national security implications. One need only witness the more than 2,000 individuals fired from various public-sector positions for suspicion of having taken part in protests and worker strikes in February and March 2011. This mass termination of mainly Shi'a employees and beneficiaries extended, *inter alia*, to government agencies, state-owned companies, hospitals, schools, sports clubs, and university scholarship-holders. The punitive response was so sweeping, in fact, that it prompted a high-profile labor rights complaint against the Bahraini government by the U.S.-based AFL-CIO, contending that the firings and suspensions violated the country's free trade agreement with the United States.[68]

As for the second half of the classic formula for political buy-off in the rentier state, the so-called "taxation effect" whereby untaxed citizens are left with no objective—read: economic—basis for political participation, it is equally clear that this individual-level link must be qualified by important conditionalities. The first problem with this line of reasoning is that, historically speaking, it is simply inaccurate. Indeed, prior to the discovery of oil in Bahrain taxes were levied on citizens, that is, on Shi'a citizens, and there certainly was no expectation of political benefits in return. Of the various forms of tax and tribute collected by the pre-oil state, the most prominent were a poll tax, a water tax for irrigation, and some claim a tax for organizing Shi'a processions during Ashura.

The former two types, Khuri explains, "were collected only from the Shi'a, on the grounds that they did not serve in the military. It should be added that they were not invited to do so."[69] On the other hand, he continues, "towns, such as al-Hidd and Rifa, where the tribal allies of Al-Khalifa lived, were not taxed. Highly placed, rich merchants did not pay taxes; they presented 'gifts,' delivered to the ruler in person, to his intimates, or sometimes to his foreign guests."[70] For Bahraini Shi'a, the taxes and tax collectors were so hated that when they were poised to be abolished as part of sweeping institutional reforms initiated by the British in the early 1920s, Shi'a seized the opportunity to express their frustration with the onerous burden, voicing strong support for the reforms and formally petitioning for British protection against the excesses of their tribal governors.[71] The result was to prompt island-wide rioting and attacks upon Shi'a villages by Sunni tribes, including by the Khawalid branch of the Al Khalifa, and ultimately the forced abdication of ruler Sh. 'Isa bin 'Ali, who opposed the reforms, in favor of his more conciliatory son Hamad.[72] Khuri tells how many a tax collector found it necessary after the reforms to relocate from the Shi'a villages to the city to escape as they said the "burden of the past."[73]

From a historical standpoint, therefore, the idea that taxation in Bahrain or the other Arab Gulf monarchies should necessarily have some relation to political rights or benefits, that they would inevitably follow the same state-building pattern exhibited by western Europe, is a dubious extrapolation. In fact, if one is impressed by anything from Khuri's account of taxation in pre-oil Bahrain, it is the degree to which the island's politics seem not to have fundamentally changed since the time of the Al Khalifa arrival: the Shi'a, whether to preserve the elite status of the regime's Sunni tribal allies or out of sheer mistrust, were systematically excluded from the nascent state apparatus, including most notably the military.[74] In return, the Baharna not only bore the economic burden of subsistence living on feudal estates, but were forced in addition to pay tribute to the Al Khalifa and their allies, who administered their fiefdoms as local sovereigns.

It was thus on the basis that they were the victims of discrimination and economic exploitation, not because they connected taxation with political privileges, that the Shi'a came out so strongly in favor of the British administrative reforms. Then as now, the primary political conflict involved a Sunni, mainly tribal elite benefiting disproportionately from the status quo, and a disenfranchised Shi'a outgroup whose members could hope to prosper far less. The case of Bahrain is therefore unlike that of other pre-oil Arab Gulf states, in which the political sphere was contested by an economically powerful merchant class and politically dominant ruling class. Crystal, in her well-known studies of Qatar and Kuwait, concludes that in both instances a formidable merchant class was content to relinquish its claim to rule in return for non-taxation and an economic monopoly in non-oil sectors of the rentier state.[75] The Shi'a of Bahrain, of course, have made no such concession, not after the end of taxation and the feudal estate system nor following the new economic opportunities afforded by the post-oil economy. At least in Bahrain, taxation and demands for "representation" have been inversely related.

One arrives, then, at the other, more basic problem with the taxation thesis: theoretically it conflates two distinct matters, namely the motivations of governments and the motivations of citizens. So while Vandewalle's famous rentier principle of "no representation without taxation" may be able to explain the conditions under which governments are less likely to demand "taxation," it says little about when citizens are likely to demand "representation" beyond ruling out a single possibility, that is, when they wish to have a say over how their taxed income is spent by the state. Under what circumstances and to what extent individuals might be spurred politically by some non-economic cause, one is left to wonder. In the end, therefore, that untaxed citizens are, *ceteris paribus*, less likely than taxed citizens to insist on government accountability is not a model of how politics operates in rent-dependent states, but a model of how it does *not* operate, and one that makes all the more baffling the current push among Bahrainis and other ordinary Gulf Arabs, untaxed as they are, for a greater role in political decision making. As expressed by Bahraini parliamentarian 'Isa Abu al-Fath,

> Nowadays in Bahrain . . . everyone is worried about politics—too much about politics and not enough about their own business. I go to the dentist, or a doctor comes to my *majlis* . . ., and the first thing he does is starts to ask me what I think about some political issue. I tell him, "Worry about your patients, and leave the politics up to politicians." But no one minds their own business anymore. It is like this now in all the GCC countries, whereas three years ago it was never like this. Even in Saudi [Arabia] they are talking politics—three years ago you would never hear that.

This revealing response was elicited by a direct question about whether an interest in political participation would exist among Bahraini and Gulf citizens irrespective of their economic situation. While Abu al-Fath skirts around the root cause of the surge in political awareness that he describes, Samy Qambar of al-Manbar does not. Even if everyone were rich, he begins in response to the same question,

> I think Bahrain would still face the issue of how things should be divided within the society between Shi'a and Sunna. When you asked at the beginning what is the biggest issue facing Bahrain, this is one of the biggest issues. The Shi'a feel they have a right to power and influence in the society, and a role in the government. This is especially so since the Iranian Revolution. The influence of the Shi'a in Iraq and Iran is very great in Bahrain, and the country needs to know how to deal with and cooperate with them.

Shi'a inside and outside the official channels of politics echo this view. The deputy head of al-Wifaq, Khalil al-Marzuq, explains:

> If the economic situation were better in Bahrain—or at least equal between the Sunna and Shi'a—the sectarian problem would become less but still wouldn't disappear altogether. This is because sectarianism has become part of the national or individual consciousness here in Bahrain since the Iraq war brought empowerment to the Shi'a there. Even post-Iranian Revolution the sectarian thinking reached a certain height, but it was never this bad.

'Abd al-Hadi al-Khawajah, founder of the Bahrain Centre for Human Rights and prominent regime critic then only recently released from prison for a fiery oration at the height of Ashura (and today serving a life sentence for his alleged part in the February 14th uprising), tells a similar story:

> Before I left Bahrain [for exile in Denmark] when I was 17, . . . there was no "sectarian problem." The political conflict at that time [i.e., the 1960s and 1970s] was between socialist groups among each other and with the government. Then

following the Iranian Revolution and more recently the Shiʻa empowerment in Iraq, the feeling in Bahrain is that they should not be marginalized anymore in the face of a ruling Sunni minority.

Yet, while the fervor surrounding Arab nationalism may have seemed to overshadow Bahrain's "sectarian problem" for a time, still it is clear from accounts of the period that animosity still burnt brightly between the two sides even prior to the upheavals in Iran and Iraq. Al-Rumaihi, writing in 1976, says of the Shiʻa of his native Bahrain,

> [their] beliefs, whilst strongly held, are at variance with the interpretation of Islamic teaching according to the orthodox sect of Islam, the Sunna, who in Bahrain refer to the Shia as Rafidi ('the Rejectors'[76]). Both points of view are fanatically held by their proponents and these differences of interpretation created the tensions which led to social and political conflict.[77]

Herein, then, lies the basic trouble with the "no taxation, so no representation" thesis, and indeed with the extant rentier state paradigm more generally. Without ever saying such explicitly, it purports to understand not only *why* people become interested in politics, but moreover their normative political orientation. Whereas, in fact, we know quite little about the determinants of individual political views and behavior in the context of the Arab Gulf, or how such behavior might be influenced by country-level variables such as the magnitude of resource wealth, ethno-religious fragmentation, or the specific strategy of political rulership adopted by a state. To do so would require, in the first place, that one actually ask individuals. And to gain access to ordinary Gulf citizens one faces many practical and political barriers; to gain access to hundreds or thousands spread across an entire nation one meets even more obstacles; and to ask them how they feel about their ruling families one should keep one's suitcase and passport at the ready.

Even absent a systematic empirical analysis of mass political orientations, however, which we shall reserve for chapter 5, it is clear based on the preceding examination of Sunni-Shiʻi conflict in Bahrain that we do know at least one thing about the bases of political action and opinion in the Arab Gulf: namely that they are not limited to the economic. On the contrary, from the nativist claims of the Baharna and their fear of falling victim to Sunni-sponsored demographic engineering, to Sunni suspicions of an Iranian-backed "Shiʻa revival" that threatens to overrun the entire region, political calculation in this Gulf rentier state is not dictated solely by economic self-interest.

Instead, the material rewards citizens might expect for remaining quiet are insufficient to deter them from seeking an active role in political life, which they do not pursue as individual benefit-maximizers but as members of a larger coalition seeking influence and control over the state itself, and this precisely over against the rival group. Ultimately, therefore, non-taxation as a means of political pressure-relief for rentier regimes is, like that of public employment, vulnerable to one fatal circumstance: when strictly economic concern is not the fundamental driver of political action; when the wealthy, corpulent, and politically disinterested "oil sheikh," the standard caricature of the Gulf Arab both within the Middle East and in the West, does not accurately represent the average citizen—who is neither rich nor poor, is politically agitated, and, above all, is either a Sunni or a Shiʻi.

Yet there remains, in addition to the group-based political mobilization described thus far, a final source of political inspiration in the Arab Gulf state that until now has gone untreated: that of religion—of Islam—itself. To this point, the line of argument has not appealed to anything intrinsic about the two conflicting groups themselves; the latter could just have well been left-handers competing with right-handers or Tamil-speakers with Sinhalese-speakers. But in fact the Sunni-Shiʻi conflict in Bahrain and across the Arab world more generally is not a group conflict simply, but one that overlaps with a 1,300 year-old religious schism precipitated itself by a dispute over political succession, a division that, as such, provides ample historical fodder for those looking to marshal adherents for a political cause.

This religious dimension is foreshadowed in al-Rumaihi's preceding description of Bahrain's social problems as stemming primarily from the Baharna's "rejection" of orthodox Islam. Elsewhere he makes the point even more explicitly, saying, "The root cause of the problem [is] the conquest of the Shia by the Sunna tribes of the mainland. The latter regarded Shiʻism as a form of heresy, and consequently missed no opportunity to oppress the original Shia inhabitants."[78] We may, of course, doubt whether the social or political outcome of the Al Khalifa capture of Bahrain would have differed qualitatively had the native population been Sunni; yet, as to the role of religion per se in stoking the flames of political conflict, there is no question.

3 Religion and Politics in Bahrain

In BAHRAIN ONE may readily distinguish Sunni from Shi'i from any number of details: speech and accent (the former pronounce the Arabic *kaf* as the English *k*, e.g., the latter as *ch*[1]); facial hair and dress (Salafis keep unkempt, often henna-dyed beards, while Shi'a are less likely to wear the typical Gulf Arab head-dress); given (Husain versus Khalifa) and, if all else fails, family name. Yet among the most straightforward methods is to observe the unmistakable adornment of private property.[2] Shi'a houses, clustered together in tight formation, fly black or multicolored flags bearing the name of the Imam Husain and other religious figures, eulogizing, "O Husain! O Martyr!" Sunni houses, often with gated entrances and garden courtyards, fly the red and white national flag of Bahrain.

Vehicles driven by Shi'a are decorated invariably with an embossed sticker decal bearing the words "God bless Muhammad and the House of Muhammad."[3] This line, with which they conclude each prayer and whose invocation of the family of the Prophet flies in direct defiance of Sunni practice, reiterates that they are indeed the Shi'a: *shī'atu 'alī*, or "the partisans of 'Ali" and the hereditary line of the Prophet against rival claimants to the Islamic caliphate.[4] For their part, Sunnis don their vehicles with the familiar Muslim profession of faith and first pillar of Sunni Islam, the *shahāda* bearing witness that "There is no God but God, and Muhammad is God's Messenger."

During the holy month of Muharram, however, in particular during the first ten days building up to Ashura proper, this religious ornament reaches a new height, crossing the line from private to public and hence drawing the ire of many Sunnis for whom such advertisement represents unnecessary embellishment and even provocation. Black banners with brightly colored, intricately embroidered calligraphy, usually in the Persian-style *nasta'līq* script, sprawl across the streets of Shi'a neighborhoods, recounting the martyrdom of Husain and brought to Bahrain, I was told, from Iraqi makers in Karbala itself. So too hang building-size portraits of local religious figures such as Sh. 'Isa Qasim, Bahrain's ranking *marja'*, or source of religious emulation for Shi'a, as well as decidedly non-local ones like Ayatollahs Khomeini, Khamene'i, al-Sistani, and Fadlallah, not to mention Hasan Nasrallah. To commemorate the occasion one may purchase a Hizballah flag, Khomeini t-shirt (which during my visit in 2008 were sold out after the first night), or for the younger revolutionary even a Khomeini jigsaw puzzle. Such overt symbolism does not go unnoticed.

Following the 2007 festivities, which saw the brief arrest of three opposition leaders for anti-government speeches, Bahrain's minister of interior, Sh. Rashid bin 'Abdallah Al Khalifa, spoke out against the "politicization" of the Ashura ceremonies, which he said had been "used to excite people through spreading false rumours, inciting hatred, belittling national achievements and seeking to erode unity." The occasion, he continued, "was also used to put up negatively worded banners and posters and flags that indicated a lack of national loyalty and allegiance."[5] Indeed, one "negatively worded banner" that particularly incensed the government and Sunnis alike had appeared the previous year under the sponsorship of the Islamic Enlightenment Society, the front of Iraqi al-Da'wa in Bahrain.[6]

A supposed quotation from a sermon by the aforementioned Sh. 'Isa Qasim, the large banner, distributed across various parts of Manama, recalled the very historical event behind Ashura itself: the decisive Battle of Karbala of 680 CE, in which the Prophet's grandson Husain ibn 'Ali, along with much of his family and supporters, were massacred by a military detachment sent by the second 'Umayyad caliph Yazid I. Our ubiquitous commentator Sh. Jasim al-Sa'idi publically denounced the banner as "a flagrant call to sectarian division in Bahrain." A columnist for the hard-line daily *Al-Watan*, close to the Royal Court, called it "a blatant violation of the constitution and a shocking incitement to sectarianism taking place months before historic elections in Bahrain and at a critical time when the region is dealing with the Iranian nuclear crisis." The message, she warned, "is an attempt to provoke Sunnis into a counter-reaction that could lead to a dangerous situation." The banner read:

> The Battle of Karbala is still going on between the two sides in the present and in the future. It is being held within the soul, at home and in all areas of life and society. People will remain divided and they are either in the Hussain camp or in the Yazid camp. So choose your camp.

For the nation's Sunnis, this "flagrant," "blatant" provocation seemed nothing short of, as the *Al-Watan* writer put it, a "declaration of war by calling upon Bahrainis to choose between the Sunni camp and the Shiite camp."[7] Its timing, moreover, coinciding as it did with heightened domestic and international tensions, was in their view either very inopportune or downright suspicious. Whatever the case, it evidenced at a minimum "a lack of national loyalty and allegiance" by those who would subscribe to such Manichean thinking. Yet, notwithstanding the rawness of its expression, a contrast to the Sunni tendency toward euphemism ("some groups," "certain people") when discussing the intercommunal conflict, the banner does little more than paraphrase the fundamental lesson, past or present, of Ashura: there exists in the world just rule and unjust rule, and it is incumbent upon the lovers of the good and the just to resist the evil oppressors, even if that means by material and bodily sacrifice. For just as the Imam Husain gave his life before he would give allegiance to Yazid, so too must all who are subjugated be prepared to forfeit earthly enjoyment for the true reward in the hereafter. In the words of the aforementioned Shiʿi activist and theologian Sh. ʿAbd al-Wahhab Husain, "The history of Shiʿism is the history of opposition against Sunni powers."[8]

Here as elsewhere one is indebted to Khuri, who provides a vivid account of the Bahraini adaptation of the Ashura ritual, the particulars of which, as he notes, "need not and do not correspond to the facts of history." He tells,

> The ritual begins on the first day of Muharram and ends on the thirteenth, reaching its climax on the tenth, the day Imam Husain was slain by the Omayyad troops. Between the first day and the sixth, the mullahs [preachers] relate Husain's military expedition against Yazid from his starting point at Medina until he arrived in Karbala by way of Mecca. They prepare the audience for the battle, which, according to the ritual, comes on the seventh day. In these six days the mullahs expound on the uncompromising stand of Husain on matters of principle. This refers specifically to his right to the caliphate, according to Shiʿa traditions. The mullahs refer to the many temptations for Husain to abandon his cause, temptations he utterly rejected. They believe Husain was chosen to be a martyr; he knew in advance that he was "destined" to lose the battle and be slain at Karbala. His martyrdom was meant to demonstrate to the faithful that "giving away one's blood for a right is an act of eternal justice," as one mullah put it. The determination of Husain to fight in spite of the temptations not to do so or of his prior knowledge of the fateful result are strongly projected in the ritual against the vulnerability of human-kind, who easily fall victim to temptations and mundane matters: material gain, positions of power, worldly pleasures, the fear of loss of wealth.
>
> Although it revolves around the person of Husain, the ritual of Ashura is depicted as a form of group sacrifice, the catastrophe of an entire family—men, women, and children. Only the infant Zain al-Abidin survived the battle; the men and children were slain as martyrs, the women were taken captive. . . . Of the many male relations of Husain (about seventeen) who took part in the battle, only three receive elaborate treatment in the ritual. . . . In the historic battle of Karbala these men were all slain in one day, the tenth of Muharram, but in the ritual each is assigned a specific day.[9]

The preachers who relate these events, which they do in *mawātim* designed specifically for the purpose[10] and fitted with enormous loudspeakers, Khuri divides into two types: those who focus on the accepted "historical" events of the battle rather than alter the narrative "to accommodate . . . the rising sociopolitical circumstances of the day"; and "those who take the battle as a symbol signifying the right of rebellion against injustices, wherever and whatever they be."[11] The latter, as one might expect, he tells have been better represented in times of political turmoil, as they were when I attended in 2008–2009. In fact, it was difficult to perceive many of the former category. For both types, though, the goal is the same: to arouse unrestrained grief in one's listeners, if only temporarily. As Khuri says, "Public opinion asserts that a mullah who cannot make his audience cry is 'no good.' [But as] the ritual comes to an end, and it often lasts about an hour, those who have been shedding real tears quickly shift back to ordinary moods." Crying at the death of the martyrs, it is held, "'assures the faithful of a place in paradise.'"

With the commemoration of the murder of Husain's stepbrother al-Qasim on the eighth night of the ritual, however, there commences an even more emotionally and politically charged feature of Ashura: mass street processions in which organized groups of mostly young men march in unison, beating their chests rhythmically and chanting religious poems that glorify the family of the Prophet (*ahl al-bayt*) and the Shi'a martyrs. Known locally as *'azza'*, the processions are led by a eulogist who composes and recites the chants, the most famous of whom in Bahrain is Sh. Husain al-Akraf, who played a central role in the Shi'a uprising of the 1990s by developing new chants "in which he connected the drama of Karbala and that of the Bahraini martyrs, Husein's fight against Yazid and the Bahrainis' fight against the Al-Khalifa."[12] For this he was imprisoned for five years, released only after the general amnesty of 2001. Among al-Akraf's poems, many of which are available online complete with video footage from Bahraini Ashura celebrations, are "Liar! O [Bahraini] Law!," "Where's Saddam?," a gibe at the late Iraqi dictator and warning to the Al Khalifa, and "Oh How You Oppressed Us!," which has been viewed more than 300,000 times on YouTube since 2007 and carries the subtitle, "God Help the Bahrainis and God Damn the Al Khalifa."[13]

On the tenth day, the day of Husain's murder, Ashura reaches its climax. It is this day with which outside observers are most familiar for its gruesome images of self-flagellation often broadcast in the Western media. This practice of *taṭbīr*, called "*ḥaydar*"[14] in Bahrain, is performed only by the most enthusiastic of the cortèges, Sunnis condemning it outright and the Shi'a themselves divided between those who deem it (or, rather, whose *marāji'* have ruled it) forbidden, permitted, or even obligatory.[15] We turn again to Khuri:

> One procession advances at the sound of drums with the participants beating their back with bundles of wire (*sangal*), or with chains whose ends are tied to sharp bits of steel that continually make slight wounds around the waist, gradually biting into the outer layer of the skin. The members of another procession, wearing white robes, beat their closely shaved heads with swords, chanting rhythmically, "Haidar, Haidar . . .," referring to Imam Ali. The blood that splatters over their bodies is intended to illustrate the horror of life when injustice prevails in the world. . . .
>
> Between the processions there march a number of separate small groups, each depicting a particular scene of the battle. These scenes include stray horses or camels covered with sheets of green and black cloth, indicating that their murdered knights belonged to the House of Ali; or huge paintings of Husain being slain by al-Shimr or grasping his infant son to protect him from the

enemy; or a young child in grief mounted on a horse treading lonely on earth, in reference to Zain al-Abidin, the only male child to survive the battle. . . .

On the tenth day these processions start early in the morning, about eight o'clock, and continue until one or two in the afternoon. The line of participants in Manama, when I observed them in 1975, continued about four hours. When the processions end, each wounded participant retreats to his own "funeral house" to wash his blood away by "Husain's water," believed to heal the wounds instantaneously. After washing their wounds the participants are offered a free meal, called *'aish al-husain* (literally the rice of Husain), to which other people are invited.[16]

Thus far we have limited our consideration to the political symbolism of the formal, ritualistic aspects of Ashura, whether the elaborate decoration that spills into public space, the mullahs' recitation of the Battle of Karbala, street processions and passion plays, or the performance of *'azzā'* and *ḥaydar*. Yet there remains another, more strictly political side of the commemoration in which political rather than religious leaders take the opportunity to address their constituencies, aided by the overflowing emotion and sense of eternal betrayal and injustice stirred up over the course of these thirteen days. It is here that the usual dynamic of Ashura is reversed, and instead of the religious making use of the political to reinforce its spiritual lessons, the political makes use of the religious—and to good effect.

Of course, not everyone uses the occasion to "excite people," "incite hatred," "belittle national achievements," or "seek to erode unity," as the state would accuse. One session I attended styled itself a forum for interfaith dialogue, bringing together Sunni and Shiʻi imams along with an Orthodox Christian bishop to discuss, respectively, the similarity of the Prophet Muhammad, the Imam Husain, and Jesus, who were said to share in common the venerable qualities of justice, self-sacrifice, divine guidance, and so on. At the same time, however, it was difficult to overlook the enormous television screen positioned directly above the tent where the discussion was being held, tuned conspicuously to an Ashura address by Hasan Nasrallah broadcast live on Hizballah's pan-Shiʻa satellite television station Al-Manar. In it he cursed the despicable, Yazid-like Israelis for their military offensive then ongoing in Gaza, reminding one that for all the efforts at spiritual reconciliation, the realities of domestic and regional politics were never far away.

For those looking to make a real political statement, the venue of choice is the early morning of the tenth of Muharram, in the wake of the almost hysterical mourning at the death of Husain earlier that night and preceding the much-awaited performance of *ḥaydar* later on after sunrise.[17] Since the mid-1990s it is an anomaly if at least one political activist is not arrested for an ardent anti-government speech at this the zenith of Ashura and of the entire month of Muharram. The year I attended the outcome would be no different. The keynote speaker was rumored to be 'Abd al-Hadi al-Khawajah, who shortly before 2:30 a.m. duly arrived outside his namesake mosque in the Manama Suq district.

Despite his being from a prominent Shi'i family that gives its name to the large and beautifully adorned mosque and attendant *ma'tam*, al-Khawajah makes no claim to religious authority, his popular following mainly a result of his well-known foundational role with the Bahrain Centre for Human Rights and, even more so, for a brazen 2004 verbal assault on the country's prime minister, Prince Khalifa bin Salman. The uncle of King Hamad, he has held the position since before independence from Britain in 1971, and among Shi'a (and no few Sunnis) is undoubtedly the most hated and feared man in Bahrain, his name rarely uttered, certainly not in public. The longest-serving prime minister in the world, Prince Khalifa was, alongside his brother the late emir, effective co-ruler from 1970 until the latter's death in 1999, and he continues to wield almost unfettered power. Al-Khawajah's unheard-of attack landed him in prison for one year.[18] In fact, far from a political asset, what little religious affiliation al-Khawajah does have, as a well-known adherent of the minority Shirazi faction of Shi'ism, is in Bahrain rather a liability. It is a testament, then, to his political rather than religious cache that he was able still to command such a general audience as the one that convened on this unusually frigid January night to hear him speak.

The title of al-Khawajah's address, whose text would quickly circulate around opposition websites along with video capturing much of the event,[19] was "How the Sacrifice of al-Husain Exposed 'the Ruling Gang' and Toppled It from Power."[20] It began by invoking the "anniversary of the martyrdom of al-Husain, son of the Prophet's daughter," and "the anniversary of the Battle of Ashura, wherein the corrupt Umayyad regime carried out the murder of al-Husain and his companions from the House of the Prophet Muhammad." "On this great occasion," he appealed to "all who are free"—"from every stream or sect," "from any social class, whether rich or poor," to "men, women, and the elderly"—he called upon them all as he called upon himself, to "stand together, to demand reform, to support what is right, to promote virtue and prevent vice, all in the name of the martyr al-Husain ibn 'Ali."[21]

He beseeched his listeners "to disengage psychologically from the unjust regime, and to refuse to give it allegiance or to allow it to rule on the necks of the people," "to break promises . . . and humiliate the people, to employ mercenaries [brought in] from everywhere in order to impose itself on the necks of [its] subjects." For "when the orders came from Yazid bin Ma'awiyah to his governor in Medina," he continued, "that he should take an oath from al-Husain or else lop off his head, al-Husain proclaimed his political disobedience and refused to swear allegiance, and [instead] prepared himself for his own sacrifice, and for that of his family." And this political defiance, al-Khawajah said, was not aimed at the person of the Umayyad ruler, Yazid, "but at the entire Umayyad regime. So when al-Husain addressed the enemy's army he referred to them, saying, 'O! Partisans of Al Sufyan!'[22] and did not say 'partisans of Yazid.'" Accordingly, the introduction concluded, "the result of the sufferings of al-Husain in the Battle of Karbala was the fall of the Umayyad Empire, a regime that would last no longer than 90 [more] years, inundated by [Shiʿa-led] revolutions brought on by the Movement of al-Husain."

The next section of the speech, titled "Sectarian Alignment and Political Alignment," cautions listeners against assuming they are part of the solution, participants in the Movement of al-Husain, simply because they happen to be Shiʿa. "Know," he said, "that the Shiʿa of al-Husain's Movement are they who stood by him and supported him against political and social injustice, and not all those who identified with *ahl al-bayt* historically or doctrinally or psychologically": "for you may be of the Jaʿafari sect doctrinally speaking, or of Twelverism ideologically speaking, but at the same time you might be one of the partisans (*shīʿah*) of Al Sufyan, or of any ruling gang who enslaves [its] people and sheds [their] blood." Thus, he warned in language that mirrored almost exactly that of the controversial Ashura banner treated earlier,

> The differentiation of people in our society today between Husainis (*ḥusayni-yyīn*) and Yazidis (*yazīdiyyīn*) is not based on the sect inherited from [their] fathers and grandfathers, nor the school of jurisprudence they rely on in their individual worship, but rather on [their] political and social stance embodied by the promotion of virtue and prevention of vice: commitment to those who are true and good, and repudiation of oppressors and the people of vice.
>
> For ordinary people in their dealings with any ruling gang are of two types: there is the one who puts principle and values first but perhaps is involved with the oppressor in earning a living or in his political and social activity; yet there is on the other hand the one who puts his own self-interest first, even at the expense of what is right and the interests of the people. And each of them will reveal his true nature when the injustice . . . and the bloodshed becomes too much, and then he either will be of the Shiʿa of al-Husain in his opinions and sacrifices, or he will be of the Shiʿa of Al Sufyan. And so a battle like that of Karbala is necessary to reveal every human [type], in front of himself and in front of others.

With this statement of what might be called the thesis of the entire address, ʿAbd al-Hadi moved on to his longest and most substantive section: "The Ruling Gang and the Necessity of Uprooting it from Power Whatever the Cost in Effort and Sacrifices." Here the subject "the ruling gang" transitioned naturally from the corrupt Umayyad dynasty, in which the right to rule "moves within one family from father to son, and which looted booty and lands, and which made God's wealth [i.e., natural resources] into a state, and enslaved the people"—all this he equated to the contemporary Al Khalifa "ruling gang" that plunders Bahrain and which claims to rule on the same basis of hereditary succession.

Neither state, he said, "was founded around a single person but rather around a gang bound by tribal or familial *aṣabiyya*,[23] [one] that uses bribery and intimidation to gain support and allegiance from the self-interested," then, this support secured, "dominates [its] subjects by force." This is why, he continued, the Imam Husain "left Medina and then Mecca fearful because he refused the political oath [of Yazid]," and left with "no supporter and no certainty . . . was murdered, and the women from *ahl al-bayt* taken captive." A regime such as this, he concluded, "chose not to accept conciliation and compromise, and thus there is no use but to uproot [it]: and al-Husain's own sacrifices as well as those of his family were the means of uprooting that state, of overthrowing the gang running it, even if [it took some time]."

He arrived finally at what the listeners had been anticipating the entire night. "The ruling gang in Bahrain," he boomed, "is embodied in the Supreme Defense Council comprised of fourteen of the elites from the ruling family, and they are: the king, the crown prince, the prime minister, the royal court minister, and others of the top ministers and officials" from the ruling family. Among them, he said, "there are not any [ordinary Bahrainis] from the Sunna or the Shi'a, as they don't trust anyone but themselves. And since the establishment of this council there have issued from it all of the conspiracies hatched against the people." All of these "conspiracies" we need not revisit at length. Suffice it to say that al-Khawajah was careful not to omit any of them: the appropriation and gifting of lands (especially reclaimed seaside lands) by the Al Khalifa; al-Bandar's report and "the strategy of sectarian cleansing" that it revealed, including of course the related program of political naturalization; the use of "tens of thousands of mercenaries from various [countries]" that "violate the sanctity of our homes and of our mosques"; and abuses of human rights and the use of torture in dealing with political activists, among whom he named one who had been recently killed in a confrontation with riot police.

For all such offenses and humiliations perpetrated by the ruling Al Khalifa gang, he directed, "the primary order must be to bring it down from power by all means of peaceful civil resistance, and by the willingness to suffer sacrifices for the sake of it, just as the result of the sacrifices of al-Husain was to bring down the Umayyad gang from power." To this end, he continued, "there must be a coordination of efforts, a putting aside of sectarian and factional differences, and an avoidance of supporting the regime's institutions or participating in them." For, he said, "we are the generation of anger and sacrifice, and from our sacrifices will come a generation that assumes the responsibility of selecting the system of government that suits it, [one] far removed from injustice, corruption, and sectarian discrimination."

He ended his long oration with a poem: "When al-Hurr bin Yazid al-Riyahi[24] demanded of our Imam al-Husain to go back whence he came or else be killed—just as we [i.e., Bahraini dissidents] perhaps may be killed—al-Husain answered, saying,

> I will go on, and death is no shame for a man,
> if he sought the good and strove [*jāhid*] as a Muslim,
> consoled the righteous through himself,
> shunned disrepute, and was at odds with a criminal.
> I offer myself up—I don't wish to remain—
> to face [Yazid's] colossal host in the desert.
> For should I live I wouldn't be blamed, and should I die I wouldn't be
>   disgraced.
> It is humiliation enough for one to be forced to live."

These final words[25] were met with chants of "Let's bring down the ruling gang!" and, though more muted, "Death to Al Khalifa!"

Thus al-Khawajah's address in the early morning on the tenth of Muharram, attended by perhaps a thousand listeners from all over Bahrain, from Manama as well as the villages, appeared by all measures to be nothing short of a call to arms against the ruling Al Khalifa in the very image of Husain's rebellion that culminated in the events of 680 CE. Indeed, as one commentator says of the online video of the speech, "People, / This guy's calling for civil war. / Stupid and *harām*. / It's *harām* for a Muslim to kill his Muslim brother. / Of course, he'd go and say that they were unbelievers [*kuffār*; i.e., Sunnis]."[26]

Yet beneath this religious imagery and bombast lies a far more measured policy prescription: political and "psychological" detachment from the state, a coordinated rejection of "the regime's institutions" in both word and deed. In this sense, the "sacrifices" of which al-Khawajah speaks are, in contrast with the overall tone of the speech, quite pragmatic and modest. The "Husainis," he says, are those who "put principle and values first," even if at the practical level they might sometimes have to interact with the state in daily life. The false Shi'i, on the other hand—the Shi'i of Yazid—is he whose involvement with the government stems not from necessity but from economic or political motivations: the one "who puts his own self-interest first, even at the expense of what is right and true." In sum, to combat a regime "that uses bribery and intimidation to gain support and allegiance from the self-interested," individuals must resist the temptations of money and power, which are offered only at the expense of their ethical principles and political freedom. For the regime, as expressed to me by another prominent (and now jailed) Shi'i critic, possesses "a bait for every fish."[27]

Figure 3.1. A banner featuring Ayatollahs Khomeini and Khamene'i hangs from the Mu'min Mosque as Ashura drummers proceed through the Manama Suq in January 2009. Their sashes bear the words "lovers of al-Husain."

Figures 3.2 a and b. Dressed in red and faces painted black, members of Yazid's army carry on pikes the decapitated heads of Husain and his stepbrother, as captured members of their family, dressed in green, follow in shackles. Camels carry the green coffins of the martyrs.

Figures 3.2 a and b. Continued

Figure 3.3. Procession-goers march in front of a poster of Sh. ʻIsa Qasim. Behind, an Ashura pennant sits atop the Bahraini national flag.

The Clerics Speak: Religious Authority and Political Participation in Bahrain

When I later had the chance to speak with al-Khawajah—some four months later, that is, after his release from prison by royal pardon—he would indeed emphasize this need for ordinary Shiʻis to avoid political cooptation. For the Al Khalifa, he said, the problem is

> just one of demographics, and how that translates into politics. In a democratic system the Al Khalifa could not continue in power, so the goal is to preclude the emergence of such a system, or to co-opt enough Shiʻa so that they have an outlet for political participation without really challenging the status quo.[28]

With the commencement of King Hamad's supposed political reform project, he continued, "The government attempted to co-opt as many Shiʻa as possible but knew that some would reject the elections and parliament and pursue other means of influencing politics. So for these people the government had another tactic:

crackdown and harsh treatment." Hence, he said, his own arrest, and those of the other 178 activists alongside whom he was pardoned.[29]

The institutional manifestation of this effort at a wholesale Shi'a boycott of the state apparatus was, following its 2006 split from al-Wifaq until its effective dismemberment after the 2011 uprising, the al-Haqq Movement for Liberty and Democracy (al-Haqq). Indeed, al-Haqq's entire *raison d'être* was its continued rejection of the parliament and electoral process in the wake of al-Wifaq's decision to join in the 2006 vote. One of its main rallying cries, appropriately, was the slogan: "This isn't the parliament we asked for!" that once decorated the walls of many a Shi'i village. The movement, however, suffered from one critical organizational disadvantage compared to its rival al-Wifaq: though it enjoyed a large grassroots following as well as the charismatic leadership of Hasan al-Mushaima', a founding member of al-Wifaq and popular hero of the 1990s intifada, it could make no claim to religious authority. Al-Wifaq, on the other hand, is led politically by its well-respected secretary general Sh. 'Ali Salman, who studied in Qom from 1987 to 1992 and thereafter "assiduously frequented the circles" of late ranking *marja'* 'Abd al-Amir al-Jamri.[30] In addition, al-Wifaq enjoys the tacit support of Bahrain's two highest-ranking clerics, Sh. 'Isa Qasim and S. 'Abdallah al-Ghurayfi, the former considered to be its de facto spiritual leader.

This disadvantage would reveal itself in dramatic fashion in the run-up to the 2006 elections. Having already made the difficult decision to take part, the leaders of al-Wifaq were faced with a vocal opposition in the newly splintered al-Haqq, which was redoubling its call for a unified Shi'a boycott of the powerless and unilaterally imposed parliament. It was at this decisive moment that al-Wifaq fell back on its main asset: its claim to represent the religious line. Already backed by Bahrain's Shi'a leaders, al-Wifaq conceived the idea of obtaining the added support of Iraqi cleric Ayatollah 'Ali al-Sistani, whose role in mobilizing the Shi'a for his country's 2005 elections had been instrumental and well publicized. His intervention in the case of Bahrain, the leaders of al-Wifaq reasoned, would be equally effective, not least as it would naturally call to mind the spectacular empowerment of Iraq's Shi'a as a result of their electoral participation. So, just months before the Bahraini elections, Sh. 'Ali Salman

> declared publically that 'Ali al-Sistani was in favor of the participation and this is hence what al-Wifaq wrote on several of its leaflets. While none of the leaders of al-Wefaq dared to say that they have received a *fatwa* from 'Ali al-Sistani, the average Bahraini was nonetheless convinced that [he] had actually issued one in which he compelled his emulators to vote.[31]

In fact, al-Wifaq did not dare to invoke the word *fatwa* because what it had received from the Iraqi cleric was considerably less impressive than this. Louër tells that, according to al-Sistani's personal representative in Bahrain, "'Ali al-Sistani answered to the solicitation of al-Wifaq in the framework of a private telephone conversation between his son, Mohammed Redha al-Sistani, and a Bahraini of al-Wifaq's sphere whose name he did not mention." The conversation, moreover, "was not meant to be made public," a fact which led al-Sistani's envoy in Bahrain to compose a public communiqué only weeks before the elections clarifying that while "His Excellency S. 'Ali al-Sistani considers that participation is most appropriate (*aslah*)," "the point of view of His Excellency the Sayyid is not a *fatwa*, not a religiously legal ruling (*hukum shar'i*). It is an objective assessment (*tashkhis mawdu'i*) and anybody has the right to make his own assessment even if this leads him to boycott."[32]

Quite apart from the controversy surrounding the legal status of al-Sistani's advice, the leaders of al-Haqq were incensed that al-Wifaq would resort to such manipulative means to convince ordinary Shi'is to take part in the elections. Louër says that in an interview with al-Mushaima', he "went as far as saying that the Shias were on the verge of committing the same mistakes as the Christians by giving too much authority to the clerics."[33] In my own meeting with al-Haqq's political spokesman, 'Abd al-Jalil al-Singace, the *fatwa* episode was said to have "coerced [the Shi'a] to vote." But since that time, he insisted, "the past four years have shown the failure of al-Wifaq to deliver on its promises," as "the record of the authorities is that they will do what they want even if you participate" in the formal political process. They convinced people once, al-Singace said of al-Wifaq, "but they can't convince people now," referring to the 2010 election cycle then upcoming.[34] But when asked about the possibility of such a backlash, two-term al-Wifaq MP Jasim Husain responded confidently,

> As for a boycott, I don't think that we will have to worry too much about that. We have backing from many religious leaders in Bahrain that call on people to vote. We met with al-Sistani in Najaf and he supports it as well. There may be some people who boycott, but I don't think turnout will be a large problem.[35]

Fighting Religion with Religion

For the proponents of total disengagement with the regime, such scheming by al-Wifaq was not to be taken lying down. Instead, Shi'a opposition leaders moved to remedy their main strategic disadvantage vis-à-vis al-Wifaq by bolstering their own religious credentials; they would fight clerical authority with clerical authority. To this end, Sh. 'Abd al-Wahhab Husain, a powerful spiritual force behind the mid-1990s uprising whose activities landed him in prison until the amnesty of 2001, left his longtime leadership role with al-Haqq sometime in 2008 to organize what was initially referred to simply as "the New Movement." Later renamed the Islamic Loyalty Movement (al-Wafa') in a clear swipe at al-Wifaq—its operative term *al-wafā'*, or "loyalty," being but one letter off from *al-wifāq*, "accord"—this new opposition faction would mimic the latter in its design, placing its political activities under the direction of a well-known religious authority in Sh. 'Abd al-Jalil al-Miqdad, who, if he hardly stood up to Sh. 'Isa Qasim as few could, at least commanded a significant following in the southern region of the country, where both he and Sh. 'Abd al-Wahhab Husain resided.

As Sh. 'Abd al-Wahhab would explain at his home in the village of Nuwaidrat, whereas al-Haqq was limited to being a political movement led by "the old guard," al-Wafa' could be a "total movement"—"religious, political, and societal"—precisely because it had a "Qur'anic basis" inasmuch as it was directed by religious leaders. Wa'ad leader Ebrahim Sharif was even more direct in his description of the movement: it was explicitly designed, he said, to be "*shari'a*-compliant" so that it "will be able to counter criticism from al-Wifaq that the other Shi'a movements, like al-Haqq, have no legitimate religious basis, such as that that Sh. 'Isa Qasim gives to al-Wifaq."[36]

The motivating fear of Sh. 'Abd al-Wahhab Husain, like that pervading al-Khawajah's Ashura polemic and that revealed by al-Singace in his remark about the government's "bait for every fish," is political co-optation. Thus, until their effective dissolution in the early stages of the 2011 uprising, al-Haqq and al-Wafa' remained the only institutionalized Shi'a opposition groups based domestically that had refused to conform to the country's amended Political Associations Law of 2005, which requires all political societies to register for approval by the Ministry of Justice and Islamic Affairs.[37] During the most recent—and, as it would turn out, final—drive for general registration in preparation for the 2010 elections, leaders of the two groups continued their conspicuous defiance of this process despite being called upon personally to meet with the justice minister himself, Sh. Khalid bin 'Ali Al Khalifa.

"Sometimes," Sh. 'Abd al-Wahhab explained, "the government sends a message to [the unregistered and thus illicit] opposition societies that it's prepared to allow them to play by its rules and become co-opted. But, if they refuse, it will play without any red lines and will stop at no immoral practices" in its fight against them. By way of illustration, he claimed the Bahraini king had met "more than once" with top opposition figures, most notably al-Mushaima' in London in 2008. King Hamad, as Sh. 'Abd al-Wahhab put it, "wished to see if they were ready to talk." But the meeting, he continued, "was not for talk but for cooptation like al-Wifaq," and when al-Mushaima' refused the authorities "decided to punish him" by cracking down on opposition activities from late 2008 to early 2009, an offensive that as we have already seen ended with a mass pardon in mid-April 2009.[38] Yet this, it turned out, was only the beginning. In August 2010, al-Mushaima', al-Singace, and much of the leadership of al-Haqq and al-Wafa' were newly arrested, this time charged under Bahrain's expansive Protecting Society from Terrorist Acts legislation of 2006. Pardoned en masse in February 2011 in an attempt to quell protests, they would be rearrested only weeks later along with the rest of Bahrain's opposition leaders—including Sh. 'Abd al-Wahhab Husain, Sh. Al-Miqdad, 'Abd al-Hadi al-Khawajah, and Ebrahim Sharif—to stand trial before a closed military tribunal. All but Sharif, who received five years, were sentenced to life in prison.

In affording a shared set of historical and ethical references, not to mention physical venues for group activity, religion can play a powerful role in catalyzing political coordination. But over and above the influence of religious *symbolism* on the views and behavior of ordinary citizens is that of political *authority*. This power is of itself undefined, malleable according to the ends of those making use of it and limited only by the degree of respect afforded those who exercise it. If al-Wifaq wishes to convince its constituents of the right of voting, here we have Sh. 'Ali Salman, Sh. 'Isa Qasim, or indeed Ayatollah 'Ali al-Sistani himself. If the proponents of regime boycott should wish the opposite, here we have an entire new organization in al-Wafa' designed just for the purpose, headed by religious authorities in their own right. All of this of course is very well served by the Shi'a doctrine of the *marja'iyya*, which accords every individual the right to choose his own source of religious emulation. Is it, though, a distinctly Shi'a phenomenon, a product only of the modalities of the exercise of religious authority inherent to Shi'ism?

It would seem not. In fact, it would seem that the notorious incident of the al-Sistani *fatwa*, or non-*fatwa* as the case may be, was only following precedent set four years earlier by the Sunni groups that agreed to participate in the 2002 elections boycotted by al-Wifaq. At that time, the previously noted leader of al-Asalah, Sh. 'Adal al-Ma'awdah, "referred to Sunni religious authorities in Saudi Arabia to obtain the edict that allowed him and other Sunnis to vote and run in the elections."[39] "Entering the parliament is not a religious act," he said, "but it becomes a must when there is a need to counter probable harm": and "abandoning the stage to 'miscreants' who would enact or pass laws incompatible with religious values would amount to a passive participation in propagating evil." While one cannot be certain what exactly is implied in this "probable harm," and whether the "miscreants" al-Ma'awdah had in mind were Shi'is, leftists, liberals, women's rights advocates, or just plain non-Salafis, it is reasonable to think that the prospect of a Shi'a-controlled parliament, however far-fetched given al-Wifaq's boycott, could not have been entirely absent. Neither can one be sure that the call to Sunni participation was not the result of governmental pressure (as seen in the passage of the Sunni Family Law) in order to preserve some semblance of legitimacy for an election already spurned by more than half of the citizenry. Whatever the truth, it is clear that Shi'a are not the only Bahrainis to exploit the influence of religious authorities for specific political ends, a fact that earns a sardonic reproach even from the *Gulf News* author of the story on al-Ma'awdah, who could not help but note "the increasing significance that religious statements from foreign-based scholars are playing in Bahrain's polls." He notes mockingly:

They have become so important that many parliament hopefuls did not have the slightest hesitation to invite religious figures to deliver lectures at their campaign tents.

Suddenly Bahrain has become a favourite destination for eminent scholars who [deliver] lectures that [have] nothing to do with the candidates' electoral platforms.

Thus the reality that has long characterized the electoral politics of Bahrain: while Shiʻa opponents worry that their participation might be tantamount to state co-option, Sunnis are concerned lest their *non*-participation should allow a Shiʻa takeover. In this way the electoral dynamic has mirrored that of society's larger communal division, in that substantive questions of policy and resource allocation—the stuff of "candidates' electoral platforms"—are superseded by a more basic disagreement over the legitimacy of the actual institutions themselves, a question that turns around the confessional balance of power enshrined therein. Sunnis have seen in the extant parliamentary structure a system of elections and parties that, even if it does provide a forum for Shiʻa frustrations while doing little to improve the lot of their own community, still serves the welcome purpose of preserving the political status quo. Their participation, therefore, has not been on account of any real enthusiasm, but is essentially negative, out of the "need to counter probable harm," that is, to forestall undesired change.

However, the powerful social and political forces unleashed by the February 2011 uprising today hint at an evolution in political calculations not only among Bahraini Shiʻa but within the Sunni community as well. Though al-Wifaq succeeded for a time in convincing many Shiʻis that the material rewards of participation outweigh their moral opposition to the system—with the help of course of some well-timed if dubious religious prodding—ultimately it was the state's lethal response to popular demonstrations, rather than a lack of legislative accomplishments, that spelled the end of five years of parliamentary experiment. In late February 2011, the 18 parliamentarians of al-Wifaq resigned en masse over the deaths of several protesters, throwing the group's considerable organizational weight behind the previously decentralized movement. The state's subsequent refusal to entertain substantive political reforms, combined with a more radicalized Shiʻa population increasingly calling into question al-Wifaq's moral and political authority over the community, have worked to preclude the bloc's return to the parliament, and indeed to exert any effective control over the Shiʻi street.[40]

So too has the uprising altered the political strategy and expectations of Bahraini Sunnis. Eschewing their longtime representatives in parliament, which over the course of nine years succeeded only in impeding the opposition, Sunnis have thrust aside the timid and largely unrepresentative al-Asalah and al-Manbar in favor of more populist coalitions headed by a new cadre of charismatic leaders. While nominally pro-government, groups such as the Gathering of National Unity and its more youth-oriented offshoots maintain a more ambivalent relationship with the state, pressing for a tougher security crackdown on the opposition and a more generous economic distribution. No longer content to be mere obstructionists, Bahrain's new Sunni factions today demand their own seats at the negotiating table, complicating efforts to broker a settlement to the ongoing political impasse and, beyond that, endangering the narrative that Bahrain's reform movement is a Shi'a-specific affair born of sectarian intentions and outside prodding.

Even as the state has succeeded in averting cross-sectarian political *co-operation*, therefore, the mobilization of Sunni citizens has raised the once-unthinkable possibility of political *co-ordination* between groups representing Sunni and Shi'a society. For this reason the government has targeted for particular persecution those calling for joint action to redress shared grievances—grievances such as corruption, land exploitation and a shortage of affordable housing, unsustainable naturalization, and a lack of political accountability. Ebrahim Sharif, the jailed Sunni head of Bahrain's largest secular society, Wa'ad, was among the first opposition leaders arrested in March 2011 after delivering a forceful speech with this message. Wa'ad itself was temporarily disbanded, its headquarters raided by police and set ablaze. Other Sunnis, including a Salafi former army officer named Muhammad Al Bu Flasa, faced similar retribution for having "broken ranks" by joining the mainly Shi'a demonstrators at the Pearl Roundabout. Al Bu Flasa was arrested hours after a public address to protesters, emerging only several months later to issue a forced apology and retraction aired on state television.

Yet, several years on, secular and Islamist Sunni movements continue to press for a greater role in political life in Bahrain, including a place at government-opposition talks launched in early 2013 from which they would, prior to the uprising, surely have been excluded. Such a development evidences a complex dialectic. Even as Sunnis reject the legitimacy of Bahrain's uprising for its roots in the Shi'a-led opposition, and would deny it a place in the Arab Spring, still they too hope to benefit from the unprecedented political opening afforded by it, to exploit the opportunity to renegotiate their onerous bargain with the ruling family. For all their accusations of sectarianism and Iranian interference, then, one wonders whether Bahraini Sunnis truly would wish to see the political clock turned back.

Conclusion: Assessing the Determinants of Individual Political Behavior in Bahrain and the Rentier Arab Gulf

With reference to the case of Bahrain, the foregoing two chapters have sought to illustrate in systematic fashion the limitations of the standard conception of citizen-state relations in the Arab Gulf states. The inquiry first outlined the various non-economic political pressures acting upon Gulf citizens and proceeded to examine why popular political involvement is not easily assuaged by the pressure-releasing levers supposed to be at the disposal of the rentier state since its description some 30 years ago. In Bahrain, we observed, public employment is neutralized as a tool for political co-option because Shi'a citizens, for fears over regime security, are disproportionately excluded from sensitive branches of the civil service and all but disqualified from police and military service. Beyond foregoing potential co-optation, this exclusion adds precisely to the political grievances meant to be relieved in the first place. As Bahrain's minister for commerce and industry revealingly asserted, "Sometimes, for good reasons, you have to be careful who you employ." In this case, such Iranophobia comes at the expense of many thousands of Bahraini Shi'a who might otherwise have been made into regime clients.

We then considered the second theorized political advantage of the rentier state, namely its capacity for non-taxation, which in Bahrain we found to perform equally poorly in winning friends for the ruling family from among would-be political opponents. More precisely, we found that the case of Bahrain gives reason to question the idea that taxation and demands for political representation should go hand in hand in the Arab monarchies *à la* the institutional economics account of eighteenth-century western Europe. Historically, we noted, the pre-oil Bahraini state did levy several forms of tax, most all of them on the Shi'a, while the latter did not on that account dare to make any claim to reciprocal political privileges. Indeed, their mere support of a British-backed initiative to modernize Bahrain's century-old feudal system in the 1920s was enough to spark attacks on Shi'a villages and island-wide communal rioting. If one should point as a counterexample to Crystal's well-known study of the evolution of political authority in post-oil Qatar and Kuwait, we noted further, the difference is that in these cases the taxed party was a cohesive and prosperous merchant class that constituted a formidable political rival. Compare this to a vanquished religious-cum-political out-group indentured to labor in independent agricultural fiefdoms.

Yet quite apart from this historical objection, we continued, the argument about non-taxation is moreover unsatisfying theoretically. Rather than explain what one would like to understand—the constellation of circumstances under which citizens will tend to incline toward political involvement—it merely elaborates one specific case: namely when they demand to oversee the usage of their taxed income. And even this positive conclusion is obscured by the dual negatives contained in the usual articulation of the argument, which posits simply: people will *not* seek a role in decision making if they are *not* taxed. Accordingly we identified two alternative causes of political inclination in the Gulf state beyond economics: ascriptive group competition as well as religion itself.

The first point has been made sufficiently. As our Bahraini parliamentarian admitted, even if all citizens were rich, there would remain "the issue of how things should be divided within the society between Shi'a and Sunna," that is to say, their respective roles in governance. Yet a separate source of political inspiration, we went on, is religion itself, such inspiration being of two types. The first is inherent to Islam, to its very history, disputed succession, and resultant political-cum-religious division. In emphasizing their unrelenting partisanship of *ahl al-bayt* and the hereditary line of the Prophet, the Shi'a continue to invoke historical claims to political rule. The immortalization of these poignant episodes of political betrayal and sacrifice in annual rituals such as that of Ashura, with its evocative passion plays, street processions, and self-flagellation, ensures that no one, Sunni or Shi'i, will soon forget the lesson of Sh. 'Abd al-Wahhab Husain, that "the history of Shi'ism is the history of opposition against Sunni powers." The tension stirred up during this month of Muharram, utilized to good effect by al-Khawajah in his rant against the "ruling gang" of Al Khalifa as it was by Khomeini against the Shah, is a political springboard to rival any other. That such enduring grievances are unlikely to be remedied merely by giving Shi'a government jobs, or by agreeing not to tax them, goes without saying.

The second way that religion may be put to political service is less specific to Islam, applicable anywhere the word of religious leaders is taken as authoritative in political matters. As exemplified in the Bahraini debate over electoral boycott, the shrewd exercise of religious authority for finite political ends can be a powerful if unpredictable influence over individual opinion and behavior. We observed how there was no shortage of religious guidance and even binding edicts for anyone looking to convince the multitude of its political duty, whether that be electoral participation along with al-Wifaq; electoral boycott and total disengagement from the regime *à la* al-Khawajah, al-Haqq, and al-Wafa'; or participation by Sunnis precisely over against these Shi'a groups, out of the "need to counter probable harm" that would come by "abandoning the stage to 'miscreants.' " Just as the Bahraini regime has "a bait for every fish," as al-Singace says, so too do the country's religious leaders possess a *fatwa* to back every political stance, and they are not shy in employing them.

In short, then, the thesis that the citizens of allocative regimes will be less inclined to make political demands of their governments because they do not pay taxes is most problematic in that it makes the implicit assumption that, absent an economic one, there is no other basis upon which such demands might possibly be made. "In the end," Luciani assures us, "there is always little or no objective ground to claim that one should get more [state] benefits," so for the one unhappy with his share "the solution of manoeuvring for personal advantage within the existing setup is always superior to seeking an alliance with others in similar conditions." Of this someone may wish to inform Bahrainis, seemingly lacking in objectivity, for whom politics is no less than the exact opposite of this description. Bahrain's political contest is tied inextricably to alliances "with others in similar conditions"—though not, as Luciani had in mind, economic conditions—and this not "within the existing setup," but precisely in order to upend or defend "the existing setup."

From here, finally, one can easily see why repression[41] may work as an explanation for the lack of revolution in the Arab Gulf, but not as an explanation for political stability or again for the political behavior of individuals. As seen in the present chapter, many a citizen will be motivated by religious or political ideals such that he is willing to risk life and limb by engaging in activities in defiance of the state; in defiance of, for example, bans on "unauthorized demonstrations," as exist in Bahrain and elsewhere. As scholars of capital punishment might attest, individuals often simply are not deterred by threats of repercussions, whether economic or physical. Witness the hundreds who since 2003 have died annually in Iraq, braving attack while attending Ashura commemorations in Karbala; those Shi'a who, despite the assurance of physical reprisal by the Saudi religious police, sneak away during the Hajj pilgrimage to pray at *jannat al-baqī'*, the revered burial site of the Prophet's daughter and four Shi'i imams; and the several dozen would-be Bahraini revolutionaries killed in standoffs with well-armed security forces and even army infantry since February 14, 2011.

No one will deny that the Gulf regimes maintain incommensurably large and well-equipped militaries and intelligence services; that these are funded by rents from natural resources; and that their use on the domestic front to repress political opponents, not rarely in brutal fashion, probably equals or exceeds their use as deterrents to foreign aggression. Yet when 'Abd al-Hadi al-Khawajah, after already having spent some two decades in exile and in prison, can stand in the streets of Manama and call for the overthrow of the Al Khalifa "by all means of peaceful civil resistance, and by the willingness to suffer sacrifices for the sake of it, just as the result of the sacrifices of al-Husain was to bring down the Umayyad gang from power"—when such a one is prepared to take this action with the knowledge that arrest and probable bodily harm will not be too far away, then it is clear that "repression" as an explanator of political behavior must be weighed against the countervailing power of individuals to suffer and even embrace sacrifice for the sake of a political cause.

At the most elementary level, then, what is required to disarm the extant rentier state framework is to show that there exist specific, not uncommon circumstances under which everyday citizens will be motivated politically by something other than or in addition to their wallets. With this examination of sectarian conflict in Bahrain we have given substance to the theoretical account of chapter 1 that explains when and why one might expect this to be the case, depicting in detail the way that group coordination and countermobilization can come to overwhelm the normal rentier politics of patronage and individualized struggles for material self-interest assumed to operate universally in the Arab Gulf states. For a more stringent empirical analysis of these central claims, we turn now to the next chapter, which introduces the first-ever study of the popular political orientations of ordinary Bahraini citizens, informed by a mass survey of the country carried out by the author in early 2009.

4 Surveying Bahrain

Presenting as it does simultaneous advantages and disadvantages from both a practical and methodological standpoint, the choice of Bahrain as empirical testing ground for the study of group conflict in the rentier state requires some preliminary words. We may begin by considering the more important, methodological implications of our selection. Foremost among these is the dialectic implied in the title of the following section between Bahrain as a model case of group conflict in the rentier state—as the Platonic idea of the failed oil state—and Bahrain as a case that is simply *sui generis*.

In the former instance, we may imagine Bahrain as a contested rentier state whose internal dynamics apply in degrees to the other Gulf nations according to the extent of group division having arisen there either exogenously by chance of history or endogenously as a result of exclusionary allocative policy. The other GCC states, then, while less perfect forms than Bahrain, share its underlying potentiality and so remain of the same class of state. That Qatar thus fails to exhibit the sort of popular political agitation born of sectarian wrangling so evident in Bahrain, then, is so largely by the historical accident that its national population is both small and by Gulf standards quite homogenous. The central upshot of this interpretation is its admission that the insights gleaned from the case of Bahrain necessarily inform the study of other rentier states.

At the other extreme sits the counterargument that little or nothing is to be gained by consideration of the Bahraini case beyond insights about Bahrain, as its exceptional historical circumstances—a native Shiʻa population conquered by foreign Sunni tribes and determined ever since to reclaim lost political autonomy—render it unique, irreconcilably different from the region's other regimes and their domestic politics. While this latter view cannot be ignored, one would like to think that the truth more closely approximates the former, more sanguine interpretation, or in any case may be found somewhere near the mean. That such hope is warranted one has some positive indicators.

Bahrain: Gulf Exception?

First, because one of the chief contemporary drivers of sectarian division in the Gulf, and source of apprehension on the part of Gulf leaders, is a militarily powerful and regionally meddlesome Iran, we have reason to suppose that all the Gulf regimes should be equally affected in proportion to the size of their domestic Shiʻa populations. Insofar as Sunni citizens and leaders across the Gulf—Bahraini, Saudi, and Qatari Sunnis alike—believe they have cause to fear for the national loyalty of their domestic Shi'a populations—to the extent, as Louër says, that it is "the Sunnis who now feel under siege" from an ominous "Shiite revival"—then, again, we should expect Bahrain to be disproportionately but not uniquely subjected to the popular pressures and worries that stem from Gulf geopolitics.

And, in fact, if one considers the bilateral relations of Iran and its GCC neighbors, one will easily perceive precisely this pattern: that antagonism is greatest among countries with the largest Shi'a populations. Whereas Bahrain, Saudi Arabia, and to a lesser extent Kuwait maintain strained relations with the Islamic Republic, the other Gulf states, with the partial exception of the UAE,[1] remain on far better terms. So while there are no doubt other reasons why one might expect Iran's relations with, say, Oman and Qatar to be more cordial than those with Saudi Arabia, the discrepancy in latent potential for domestic Iranian influence as a function of a state's citizen Shi'a population is perhaps not the least important.

Moreover, evidence that a group politics of sectarian rivalry operates in the Gulf outside of Bahrain may be found simply by reflecting on the present state of Sunni-Shi'i relations across the region—both citizen-citizen relations and citizen-regime—which have reached a nadir not seen since the immediate aftermath of the Islamic Revolution. In Saudi Arabia, in Kuwait, and even in countries with small Shi'a populations and with little history of sectarian friction, such as Qatar and the United Arab Emirates, Shi'a communities are increasingly under scrutiny by Sunni citizens and rulers made nervous by a confluence of worrying domestic, regional, and international developments. This concern is aptly conveyed in an anecdote related to me by a Pakistani Balouch living in Qatar, whose brother, like many others of his ethnicity who fill the ranks of Gulf militaries, was employed in the Amiri Guard. Upon notifying his superiors of his plans to be married, the brother was informed that he should not go through with the wedding if he wished to keep his job. His prospective wife was a Qatari resident of Persian origin.

Of course, rather than evidence the universality of hatred between Sunni and Shi'i Muslims, the pervasiveness of sectarian tension today is an indication once again of the common internal and external forces acting upon Gulf citizens. Beyond the shared Sunni fear of Iranian expansionism, these forces stem largely, as examined in chapters 2 and 3, from the very policy choices of governments themselves, which have worked consciously or unconsciously to politicize group identities. Rather than attempt to downplay latent social divisions, most institutions adopted by Gulf states aim deliberately to emphasize society's latent potential for group-based political coordination; to augment the political salience of ethnic, religious, tribal, and other identities in order to help preclude the emergence of a much more dangerous sort of political movement: one based on shared policy preferences rather than shared descent.

Even absent political machinations on the part of rulers, moreover, the transmission of social group distinctions into the political arena is a natural consequence of the unequal pyramid of distribution inherent in the rentier system. The windfall profits of petroleum exports are not distributed in a way that is socially or politically agnostic, but, as Okruhlik says, on the basis of "family relations, friendship, religious branch, and regional affiliation."[2] In its biased distribution of oil rent, then, the system creates both relative winners and relative losers, and so both political friends and political enemies.

The other main trade-off occasioned by the choice of Bahrain as a research subject is at once practical and methodological. This is the consideration that, in choosing to study the ideal case of group conflict in the Gulf—the case that offers the most to observe—our conclusions may be susceptible to selection bias; to the criticism that one has, for those versed in such language, selected on the dependent variable. Distinct from the first point above, this concern is not about the Bahraini case *per se* but about the lack of one or more additional country observations to which to compare it in a structured cross-national analysis. Undertaking the latter, one could better discern which of one's country-level outcomes are attributable to the independent variable of interest—ascriptive group division—and which to the peculiarities of the individual cases themselves. Absent this, some skeptic might say, one can draw through a single point whatever regression line he likes. That is to say, of a single case any interpretation is equally defensible.

Of course, one would have liked to replicate the Bahrain study in one or all of the remaining Gulf countries, complete with respective nationally representative surveys of citizens that capture, *inter alia*, sectarian and other group affiliations. In this way, not only would one gain extra statistical leverage with which to test the individual-level relationship between group membership and political orientations and behavior, but, in addition, one could make a more robust test of the theory that country-level differences in social group configuration relate to overall regime stability in the Gulf. Yet in that case one would also have had to contend with a much-augmented set of practical challenges that in Bahrain alone were enough to prove nearly insurmountable. To be sure, the present dearth of such quantitative studies of the GCC states is not for lack of demand.

Surveying Bahrain

Even before Bahrain began systematically expelling and denying entry to academics in the wake of the 2011 uprising,[3] barriers to conducting research in the country were substantial, to say nothing of work touching on the most sensitive of social and political topics. And, in a context where the mere number of Sunnis and Shi'is in the citizenry is a veritable state secret, investigation into Bahrain's sectarian relations is barely tolerated. This fact was much in evidence in my dealings with the Bahrain Center for Studies and Research (BCSR), a now-defunct government institution that served for almost 30 years as a research clearinghouse for various ministries and, during my stay in Bahrain, as the sponsor of my Fulbright fellowship. (One imagines that my resulting association with the U.S. State Department explains why I was allowed to remain in Bahrain at all.)

Dissolved less than a month after my departure under murky circumstances,[4] the BCSR had originally agreed to administer my Bahrain survey through its own political polling unit, the latter having been established in the early 2000s with the help of several University of Michigan trainers sent to the country by a U.S. government grant. But after almost a year of meetings and edits of the survey instrument to remove "sensitive" questions—by the end there remained just 65 of the 110 or so standard Arab Barometer items—it became apparent that my fellowship funding would sooner be exhausted than the BCSR actually carry out the work. Finally, I was told that the center, a state institution, could not risk administering a survey the results, or mere fielding of which, would likely earn the government's displeasure. We arrived, at length, at a compromise whereby the BCSR would continue to sponsor my residency while I organized and carried out the survey alone.

In the meantime, I received frequent, unsolicited e-mails and calls from altruistic persons offering to be of service in carrying out my study. One message arrived from the director of a nonexistent "National Centre for Studies," another from a worker at a government-affiliated organization suggesting that if I would only send my survey data, he may be able to assist in "analyzing" them. When I thanked the latter for his offer, noting that as yet there were no data to analyze, he replied that in any event he had a "friend" who would be happy to work as a field interviewer for the project, providing a mobile number at which I might contact him. Then, when interviewing finally commenced some months later, I received a call to come that day to the local police station, where the area police chief wished to discuss the survey project.

As I stepped into his office, I saw in front of him a photocopy of the full survey instrument, which he said he had obtained from a respondent who happened to work for the National Security Agency and was apparently alarmed at some of the questions. Sensing a genuine interest on his part and hearing that he agreed there was strictly speaking nothing illegal about such a study, I asked if he might consent to fill out a questionnaire himself that I could later retrieve, seeing that no one else from the Interior Ministry had agreed to meet with me as part of the elite interview portion of my research. My calls the following week, however, would go unanswered.

If the Bahraini government is thus made nervous by such a sensitive investigation,[5] the ordinary people who must form the basis of it are no less so. In light of my own experience, of course, one certainly cannot blame them; yet, this pervasive mutual suspicion makes the task of conducting face-to-face interviews difficult. Compounding matters is what Fearon and Laitin accurately refer to as the island's "metrocommunity scale,"[6] which they argue is a primary reason why Bahrain so far has averted sectarian civil war. Some sense of this may be related by a conversation I overheard in Sana'a airport between two Yemenis bound for Bahrain. One, distinguishable by his dress as a long-time Bahrain resident, asked the other if this was his first time traveling to the country. When the latter replied that, yes, it was, the former nodded his head, saying, "I thought so. I haven't seen you there before." The response elicited a confused expression on the face of his interlocutor. "What do you mean you haven't seen me there before?" he replied. To this day I can hear the exact words of that Yemeni-Bahraini: "Bahrain is like the village: everybody knows everybody else." As one who probably remembered well from childhood what it was like to live in a tiny, remote village, the Yemeni traveler immediately understood.

Beyond mere suspicion, then, Bahrain features a general lack of social anonymity to an extent not seen elsewhere in the Arab world, and this sometimes uncomfortably so when one finds oneself outside of a handful of communally mixed, urban areas. For Bahrain, taken as a whole to be no more than a village by our Yemeni observer, in fact is home to several dozen even more isolated village enclaves settled, with very few exceptions, exclusively by Sunnis or Shi'is. Until 30 years ago this sectarian residential separation, which linguist Clive Holes describes as an "almost apartheid-like system of voluntary segregation,"[7] extended across the entire island, to urban and rural areas alike. Indeed, it is only for this extreme isolation that Holes was able to complete a comparative study of the country's Sunni Arab and Baharna dialects. On this he remarks,

> One consequence of the separation of the two communities has been the preservation, over more than two centuries, and in an area no bigger than a medium-sized English county, of a major dialectical cleavage that pervades all levels of linguistic analysis: pronunciation, word structure and vocabulary. The historical origins of this split, as is usual in cases of major communal differentiation of this kind, are geographical.

In the wake of the February 2011 uprising, which saw a labyrinthine system of security checkpoints installed by riot police, military, and even ordinary citizens worried over "infiltration" by demonstrators or regime loyalists, this communal segregation has been entrenched still further.

For one aiming to conduct personal interviews across the whole of the island, the upshot of all this is that field interviewers are immediately identifiable as, first, Bahraini or non-Bahraini; second, as Sunni or Shi'i; and perhaps, depending on the respondent's knowledge of dialect and family names, even as a resident of a particular village or region. Moreover, this information may then be translated into a perhaps stereotypical but not altogether unreliable overview of one's likely social and political affiliations. For Sunnis, for instance, one can likely glean tribal versus nontribal affiliation, and status as *mujannas* versus a non-naturalized citizen. For Shiʿa, outward characteristics can signal both religious (e.g., Shirazi vs. *vilāyat-e faqīh*) and political affiliation (pro-government vs. al-Wifaq supporter vs. follower of the more radical opposition). One called "al-Rumayhi" will at once be supposed a government ally perhaps hailing from the village of al-Jaw, the traditional home of the Al Rumayh. One whose accent betrays him to be a resident of al-Diraz will be branded a follower of Sh. ʿIsa Qasim, born in the village, and thus most likely of al-Wifaq. Consequently, such an individual may meet with a cold reception in the neighboring Shiʿa villages along al-Budayyiʿ Road.

As if not already complicated enough by this claustrophobic social atmosphere, survey work in Bahrain must also contend with the physically claustrophobic villages themselves, which are both isolated and poorly served by roads, the latter often little more than paths cut through the sand and gravel. The roads that are paved are so narrow that one is often stuck inside one's vehicle at length when, inevitably, another appears traveling in the opposite direction. At which point one is forced to navigate the same narrow road in reverse for several hundred feet. Village driving is so difficult, in fact, and one is so easily recognizable as an outsider, that U.S. government employees have long been banned from visiting. One embassy official told me with pride how he had defied this order and driven through a village adjacent to his housing compound in order to observe the colorful political graffiti about which he had been told. A resident of the Shiʿi village Karranah complained bitterly that the police themselves refuse to enter unless to make an arrest or to chase away teenagers burning tires. Even in the event of a simple car accident, he explained, exasperated, the police demand that villagers themselves drag the damaged vehicles to the main road for examination, so that any facts of the incident gained by observing the wreckage or through interviewing witnesses are necessarily lost.

As a result, many of the Shiʿa villages, though the capital and most ministry headquarters be but five miles away, have learned to operate to a startling degree independently of the state, referring disputes to local religious notables, aiding poorer residents through the local village charities, and even undertaking infrastructure repairs and construction. Being thus largely interdependent upon each other for most everything save for electricity and sanitation, it is understandable why village residents may not immediately welcome outsiders, not to mention ones asking probing political, social, and religious questions.[8]

It is also easy to see how this isolation, this isolated frustration, may easily erupt in the form of protests, tire-burning, and other localized violence, a final and most severe impediment to field research in the Bahraini villages. The riot police, generally loathe to intervene in these so-called terrorist acts out of fear for their own safety, are content to assemble in dozens of armored SUVs along a village's main access road, effectively cordoning it from the outside. In times of substantial Shiʻa-government confrontation, then, such as was the situation in early 2009 following the post-Ashura arrests of ʻAbd al-Hadi al-Khawajah, Hasan al-Mushaimaʻ, and other senior political activists, mere entry and exit is made problematic for village residents, and any interviewing is out of the question. So it was that only four months later, after King Hamad's mass pardon of April 2009, were we able to commence surveying in these areas, and even then there remained a small number of locations where, due to continuing tension or the unease of would-be field interviewers, we were unable to conduct interviews. Indeed, these locations included the very home villages of several of the Shiʻi field interviewers, who nonetheless refused to work there for fear of being deemed by their neighbors government spies.

Fortunately, such difficulties were limited to the mass survey portion of my Bahrain research; the other half, structured interviews with political leaders representing the various factions of Bahraini society, was able to progress more smoothly. The only complexity here was making the acquaintance of some of these individuals, in particular those who belong to political societies with whom the U.S. embassy deliberately maintains no ties. This included, at the time, the two main Sunni political societies—whose ideological bases in Salafism and the Muslim Brotherhood precluded such cooperation—as well as all the Shiʻa groups and movements not named al-Wifaq, presumably on the consideration that they are technically illegal. Having attended many embassy-sponsored social events and receptions, I was struck at length by the monotony bordering on perfunctoriness of the political guest lists, which were made to include invariably several (generally lower-ranking) Al Khalifa bureaucrats; a familiar set of technocrat MPs from al-Wifaq; a diverse group of former Marxists turned "liberals"; and perhaps a few Sunni "independents." The far limit of this political diversity was the presence of one particular member of al-Asalah—"a moderate," I was assured. Never did I observe anyone affiliated with the Muslim Brotherhood-affiliated al-Manbar, or with the non-parliamentary Shiʻa opposition—"the Haqqis," as U.S. embassy officials were fond of calling them.

All the same, I was able to meet a quite representative set of individuals spanning the full range of Bahrain's political continuum. These included at the time of interview: two members of parliament from al-Manbar; three from al-Wifaq; one from al-Asalah; one Salafi independent; the head of the liberal-socialist party Waʻad (not then represented in parliament); the head of the liberal Progressive Tribune Society (neither represented); the founder of the Bahrain Centre for Human Rights; a senior leader of al-Haqq; and a famous Shiʻa scholar and founder of the then recently formed "New Movement," known now as al-Wafaʼ.[9] I also attended the weekly *majālis* of several parliamentarians as well as those of prominent Sunni businessmen, who, judging by their speech, guests, and choice of television programming (the state-run Bahrain TV is a sure sign), were at least nominally pro-government if essentially seeking to appear apolitical.[10]

Given the liberal use of these interviews in service of the argument thus far, my purpose in conducting them may already be clear, namely to complement and corroborate the secondary source material to which one must unavoidably appeal in studying communal conflict in Bahrain, which, whether newspaper or website, is equally one-sided, abridged, and often shrouded in euphemism. If the responses obtained in my interviews may be assumed of a similar quality—though one will likely agree from the quotations cited thus far that few from either side seem to mince words—then at least they have been obtained first-hand and, since all were asked the same questions, are readily comparable. International outreach being, however, an explicit strategy adopted by Bahrain's sectarian-based political societies in shaping the debates vis-à-vis the government and their respective rivals, one imagines that these interviews were viewed by participants in a similar light, that is, as a chance to give a clear statement of one's political positions and ideology over against those of opposing parties.

Executing the First Mass Political Survey of Bahraini Citizens

Having thus outlined the main methodological and practical difficulties occasioned by the choice of Bahrain as a subject with which to test the thesis developed here, we may move now to a more detailed overview of the mass survey itself, data from which form the basis of the empirical analysis of the following chapter. Though I was present in Bahrain from April 2008 to June 2009, active surveying could begin only in January 2009, following the aforementioned compromise with the BCSR, and lasted until early June. This unusually elongated five-month time span was a direct result of the post-Ashura disturbances noted already, the political situation calming only in late April. Even then, many of the more embattled Shi'a villages would for some time remain unsuitable for interviewing.

On this account field interviewers were forced to begin surveying in mixed or Sunni-dominated areas first and expand to the rural Shi'a districts as they became accessible. As a rule, Shi'i fieldworkers conducted interviews in Shi'a-dominated areas, and likewise Sunni interviewers were sent to Sunni areas, though in urban centers like Manama, Hamad Town, Isa Town, and parts of Muharraq, inter-sectarian interviews were inevitable. In the end, however, this was a fortuitous development in that it embedded into the survey a natural experiment: one that revealed the effect on respondents of being interviewed by a member of the opposite confessional group rather than by a co-sectarian.[11]

The one substantive contribution of the BCSR to the execution of my survey, though an critical one to be sure, was its provision of a random sample of 500

Bahraini households that it received directly from the Central Informatics Organization, which administers the national census and maintains this and other electronic population databases. Since Bahrain's administrative units ("block numbers") correspond numerically to one of twelve geographical zones, I was able to confirm before commencing surveying that the sample was in fact reflective of the general population distribution.[12] Indeed, it even included two addresses in the remote Hawar Islands, an archipelago used mainly as a military outpost situated just a few miles off the western mainland of Qatar.

One can easily confirm the representativeness of the sample by comparing the frequency of block numbers to the known populations of the districts to which they correspond.[13] For example, the 100 and 200 blocks make up the Governorate of Muharraq, whose citizen population was officially reported in 2010 as being 102,244, or 18.0 percent of 569,399 total Bahrainis.[14] Another figure based on the number of voters registered for the 2010 parliamentary elections puts this proportion at 57,233 of a total 318,668, or again 18.0 percent.[15] Computing the proportion of 100 and 200 blocks in the survey sample, then, we see that Muharraq households make up 92 of the 500 total, or 18.4 percent. When we repeat these calculations for the remaining four governorates we find that the rest of the sample contains 83 or 16.6 percent Capital Governorate households, 145 or 30.1 percent in the Central Governorate, 150 or 30.0 percent in the Northern, and 30 or 6.0 percent in the Southern. Within the bounds of sampling error, the respective 2010 census figures are essentially identical: 11.9 percent, 29.7 percent, 34.1 percent, and 5.8 percent.

More than just geographically representative in the aggregate, moreover, the BCSR sample includes at least one respondent from almost every Bahraini village, district, and city. Indeed, there are few areas of Bahrain signified by a proper name that are not represented in the sample. Lastly, while this 500-household sample is smaller in *magnitude* than those employed elsewhere—for example, in the other Arab Barometer surveys—yet, because of Bahrain's miniscule population, the *proportion* of citizens that were interviewed is greater than any mass political survey administered to date in the Arab world outside of Qatar, at approximately 1 interview per 1,100 citizens.

As mentioned previously, however, we were unfortunately unable to complete all 500 interviews on account of the ongoing political and social turmoil that lingered even after the mass pardons of late April 2009 meant to ameliorate it. Of the full sample, therefore, 435 interviews—87 percent of the total—were completed. The unfinished areas are dominated by the crowded, urban neighborhoods of Manama (19 interviews unfinished) and Muharraq (14); and the remainder are spread across isolated Shi'a villages and suburbs and adjacent Sunni enclaves. These two types of areas proved particularly challenging in that field interviewers who were unfamiliar with the neighborhoods, especially females, were loath to go there, as one is forced to wander through foreign territory from the nearest road wide enough to fit a car; while those who did know the areas were equally unwilling to conduct interviews for fear of gaining a reputation as a spy. If the areas are thus not excluded in as random a fashion as one would have preferred, these blocks at least are spread rather equally between Sunni- and Shi'a-populated districts,[16] and even in these problem locations we often salvaged one or two interviews if sampled households happened to be located in a less isolated position. One direct upshot of this difficulty, however, was the underrepresentation of Shi'i females, which form only 29 percent of the Shi'a sub-sample.

Finally, the survey instrument itself is the standard Arab Barometer questionnaire with only slight adjustments to fit the Bahraini context. The only substantive change was the inclusion of two open-ended questions at the conclusion of the interview. The final questionnaire contained 106 separate items inclusive of demographic details, and it required approximately 35 to 45 minutes to administer. As the sample contained the exact address (house number, street number, and block number) of the sampled households, these were located using the now-defunct BahrainExplorer website, a searchable GIS map of the island maintained then by one of the BCSR's commercial subsidiaries, GEOMATEC.[17] This resource proved invaluable in directing field interviewers along Bahrain's labyrinthine and ill-marked roads to find equally ill-marked houses; without it, indeed, the exceptional and fortuitous sample would have been rendered of little use.

Despite this aide, however, an overall participation rate of about two-thirds of eligible respondents, even lower in predominantly Sunni areas, betrays the general ambivalence of ordinary Bahrainis to such a seemingly strange project, which, they reasoned, at best could do them no benefit and at worst might be no more than a ploy to discover non-allegiant subjects. It was clear that on the whole Bahrain's Shi'a saw more to gain from participation in the survey than did Sunnis, some of whom viewed the entire project with suspicion and as an elaborate, covert test of their national loyalty. One might easily have expected the opposite, that in general the Shi'a would have shown more distrust toward field interviewers given the prevailing political climate. Though some certainly did, and asked the latter if they were sure that the information would not be passed on to the government, even more Shi'i respondents asked whether the results of the survey would actually make any difference, that is, improve the lot of the Shi'a, of this or that village, or of Bahrainis generally. On the rare occasions when I was forced to accompany field interviewers to the villages to help prove that he or she was not a government agent, respondents almost always implored that their contribution be used to somehow "improve the situation." In case of refusal, ineligibility, or repeated non-response, fieldworkers moved three doors to the left of the sample-designated house, while obliging respondents 18 years and older were selected, as in other Arab Barometer surveys, using a standard Kish Table.

Bahrain's Confessional Geography

The last time the government of Bahrain reported official demographic statistics on its Sunni and Shi'i communities was in its very first census in 1941, which put the percentage of Shi'a citizens at 53 percent of the population.[18] Over the intervening 74 years, speculation about Bahrain's evolving Sunni-Shi'i balance has become both a local flashpoint and a source of frustration for those who study the country. This ambiguity has been complicated only further by the government's decade-long program of naturalizing Arab and non-Arab Sunnis for work in the police and military. Because Bahrain's Sunni-Shi'i balance is not simply a product of nature, then, there is no straightforward way to extrapolate it scientifically. As a result, a wide range of disparate estimates put Bahraini Shi'a variously at between 55 percent and 75 percent of the current citizen population, reflecting conflicting anecdotal impressions about the relative birth and immigration rates of the two confessional communities. Yet according to the direct sampling of the Bahrain mass survey, the true proportion, at least as of early 2009, is likely closer to the lower end of this continuum. Shi'i respondents made up only 57.6 percent of those surveyed, corresponding to an estimated range (based on the 95 percent confidence interval of the mean proportion) of between around 53.0 percent and 62.3 percent of the citizen population.[19] To the extent one accepts the general view that Shi'i population growth outstripped that of Sunnis over the half-century following 1941, this survey finding would seem to call attention to the pace and scope of Bahrain's modern program of Sunni naturalization.

The results of surveying are depicted visually in Map 4.1, which more fully reveals the representativeness of the 500-household sample. Here, blocks with at least one completed interview are shaded to reflect sectarian composition; those not included in the sample are dotted; and unpopulated districts are unfilled. Map 4.1 serves, in the first place, as a basic demographic map of the country, illustrating well the overall population patterns across the island. None of these is more striking than the barrenness of the southern two-thirds of Bahrain, where, apart from the lower half of Riffa; the expatriate enclaves of 'Awali and Riffa Views; a few sprawling royal palace complexes; and the Sunni seaside villages of al-Zallaq, 'Askar, al-Jaw, and al-Dur, the land is uninhabited except by police and military personnel, inmates at Bahrain's notorious al-Jaw prison, and foreign laborers housed in compounds dotting the desert landscape. The thirteen artificial islands that make up the would-be expatriate oasis of Durrat al-Bahrain at the southeastern tip of the country were unfinished at the time of surveying and in any case are unlikely to house many Bahraini nationals.

Likewise, the large swaths of unsurveyed blocks in the north of the island largely correspond to reclaimed or uncultivated land as well as commercial and industrial areas. On the island of Muharraq, for example, the large, central space

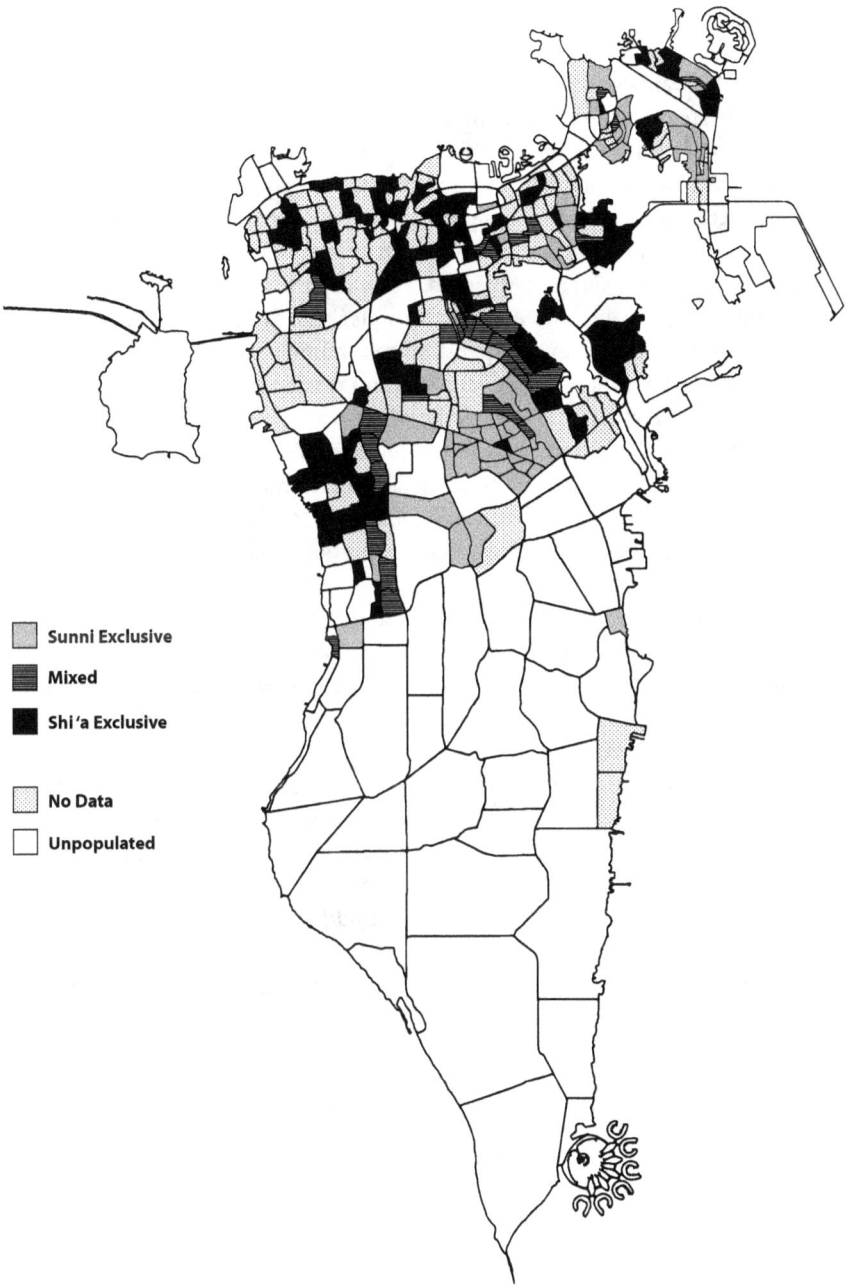

Map 4.1. Confessional map of Bahrain.

corresponds to Bahrain Airport, while the southeastern peninsula is in fact a network of dry-docks and warehouses known as the al-Hidd Industrial Area. The manmade Amwaj Islands off the northeast coast of Muharraq form yet another posh expatriate community. On the western coast one finds sparsely developed reclaimed land, although the neighborhood of Busaiteen is fast encroaching in that direction. Two separate bridges connect Muharraq to Manama. An unpopulated area stretches southward along the eastern coastline of Manama, where one finds a large district of hotels, embassies, and government buildings known as the Diplomatic Area, the National Museum, a public park, a marina, the National Library, and finally the gigantic Al-Fatih Mosque. Nearby in Juffair sits Naval Support Activity Bahrain, an ever-expanding base home to the United States Navy's Fifth Fleet. So too are there few private Bahraini residences along the entire northern coast of the Capital Governorate, the whole region being formed of reclaimed seabed and consisting, from east to west, of the newly finished, three-kilometer-long Bahrain Financial Harbor, the artificial Reef Island home to many Saudi and Western expatriates, some half-dozen shopping malls and supermarkets, a sprawling Ritz-Carlton Hotel complex with a private harbor and beach, and the centuries-old Bahrain Fort.

To the west is the Shi'a-dominated Northern Governorate, where settlements follow al-Budayyi' Highway. The road, site of many an opposition demonstration and march, straddles a dozen or so rural Shi'a villages before terminating in the Sunni enclave of al-Budayyi' situated on the far northwest coast of Bahrain. Several large, reclaimed islands being constructed to the direct north were uninhabited at the time of surveying, and much of the area surrounding al-Budayyi' village proper consists of expansive, privately owned gardens. South of al-Budayyi' the coastline is dotted with spectacular mansions and resorts, interrupted only by the King Fahd Causeway to Saudi Arabia. The large islands connected by the causeway, Umm al-Na'san and Jiddah, are privately owned by the king and prime minister, respectively.

South of the causeway entrance, the seaside compounds continue until al-Zallaq in the Southern Governorate, interrupted, as far as one can tell, only by a lone public beach. Caught between these private coastal plots and the extended western border of newly constructed Hamad Town are the rural Shi'a villages of Dumestan, Karzakan, al-Malkiya, Sadad, Shahrakan, and Dar Kulaib, among the most destitute places in Bahrain and not infrequently the sites of violent confrontations with the government. The extreme discrepancy in the apportionment of land in this western coastal region helps fuel one of Bahrain's most explosive political issues, one that unites Sunni and Shi'i alike against the perceived excess of the Al Khalifa and members of historically allied families, who appear inexplicably well-endowed of premium property while ordinary citizens are suffered to make the best of the remainder—if, that is, one is lucky enough to own property at all.

The Central Governorate,[20] finally, exhibits a similar pattern of segregated population centers. Here, though, the eastern coastline is dedicated primarily to industrial and naval use: the northern portion of the Sitra peninsula (once an island) is a warehouse zone; and the southern and eastern regions are used by the Bahrain Petroleum Company (Bapco) and other petrochemical firms for refinement, storage, and distribution. Directly south of the Bapco complex is an equally large Aluminum Bahrain (Alba) installation, which uses the adjacent coast for its own smelting operations. The only notable non-industrial users of this shoreline are the unfortunately located al-Bandar Resort and neighboring Bahrain Yacht Club.

The interior portion of the peninsula is divided between a half-dozen Shi'a villages, whose residents complain bitterly of the transformation of their landscape once known for its verdure and natural springs. One villager told how two local children had recently drowned when the beach they used for swimming was unknowingly dredged for sand, creating a precipitous drop but a few feet from shore. When asked what had been done to remove the danger, he pointed to a tattered fence bearing a "Keep Out" sign erected, he said, by the responsible company. Owing perhaps to such circumstances, and to their geographical isolation, the Shi'a of Sitra are commonly held to be among the most "extreme" in their anti-government views and xenophobia, and I was met more than once with surprise and horror upon mentioning that I would travel there to assist in surveying. Also considered part of greater Sitra are the Shi'a villages of al-'Akar, Ma'ameer, and Nuwaidrat, which straddle the highway that runs through the peninsula after it reaches the mainland.

The other mainland Central Governorate population is concentrated around several centers: the Shi'a villages of A'ali and Salmabad to the far west and northwest; the large, ethnically mixed Isa Town, whose suburbs span the length of the eastern interior coast; and the northern portion of Riffa, separated from Isa Town and the mainland Sitra villages by the six-lane Istiqlal Highway. In the western third of the Central Governorate, in the lands surrounding A'ali, are found what remains of the Dilmun Burial Mounds, a necropolis consisting of some 100,000 above-ground tombs believed to date to the fourth millennium BCE. Bifurcated unceremoniously by the Sh. Khalifa bin Salman Highway that divides the Northern and Central Governorates, these mounds also account for the large area of unsurveyed blocks situated north of Hamad Town in the Northern Governorate.

All these demographic peculiarities are summarized in Map 4.2, which shows in greater detail the confessional composition of the areas surveyed at the level of block number. For those blocks where no surveying occurred, it is, again, indicated whether this is for lack of data or for lack of a resident Bahraini population. Those blocks nominally populated by Bahrainis—the personal islands of the king and prime minister, the palace compounds around Riffa, or the large, block-sized

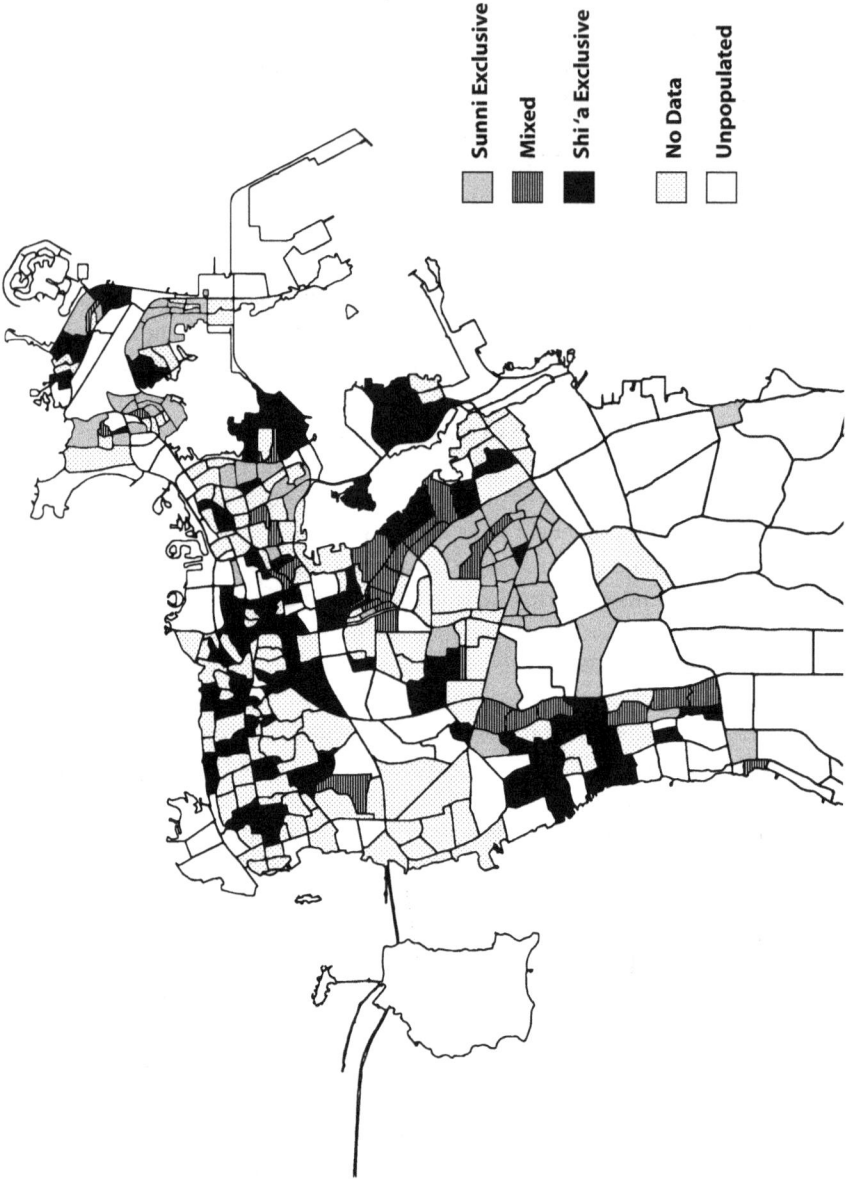

Map 4.2. Confessional map of Bahrain, detail.

Sunni Exclusive

Mixed

Shi'a Exclusive

No Data

Unpopulated

private compounds of the western shore—that in reality have no chance of appearing in the sample, are marked for our purposes as "unpopulated." If the picture that emerges be common enough knowledge to the average Bahraini, still I am aware of no one who has yet put it to paper. To my knowledge, this diagram is the first confessional map of the country. In fact, of the two demographic maps of any sort I can find of Bahrain, both of which show only rough population density, one is published by the U.S. Central Intelligence Agency and is based on the 1981 census,[21] and the other is proprietary and its source unspecified.[22]

While one might have hoped to have visited more of the "no data" blocks in order to paint a more complete portrait, this map is a by-product of the Bahrain survey rather than the goal of it. In any case, the population of many of these blocks, if not zero, is so sparse as to require a sample that is orders of magnitude larger than the 500 households surveyed in this case in order for them to appear at random. And since the government population data are reported only at the regional or governorate level of aggregation, for fear of giving further empirical ammunition to the many who decry its blatant sectarian gerrymandering of electoral districts, one is unlikely to soon have recourse to such a sample.

What one can do alternatively, however, is utilize this assumption of gerrymandered electoral districts to one's advantage to serve as a basis upon which to construct an alternative sectarian demographic map that might at once confirm and expand upon the patchwork of Maps 4.1 and 4.2. This effort is represented in Map 4.3, which assumes the communal composition of a block number according to the electoral district to which it belongs. Those districts carried by al-Wifaq in the 2010 election are assumed to have Shi'a majorities; conversely, those won by anyone else, that is, by Sunni Islamist candidates and pro-government "independents," are assumed to be Sunni majority.[23] In comparing the country-level Maps 4.1 and 4.3, then, one easily notices an astonishing correspondence in the sectarian composition of the blocks, even in the most heavily integrated areas of Isa Town and Hamad Town. This simplified electoral-based map is not perfect, of course, under-representing the Shi'a presence in Muharraq and the Sunni presence in south Hamad Town and western Manama. Yet it does offer a useful estimate of what the confessional map may have looked like had we been able to interview, say, 10,000 households, to say nothing of its rather shocking confirmation of systematic sectarian gerrymandering and disproportionate inclusion among the Sunni-majority districts of large swaths of unpopulated territory.[24] Indeed, when one does the math, one finds that the mean Shi'i district represented 9,533 electors in 2010, the average Sunni district 6,196.[25]

A Portent of Division to Come

More notable for our purposes, however, is what these illustrations say about Sunni-Shi'i segregation in Bahrain, which, if it perhaps no longer exactly fits Holes's

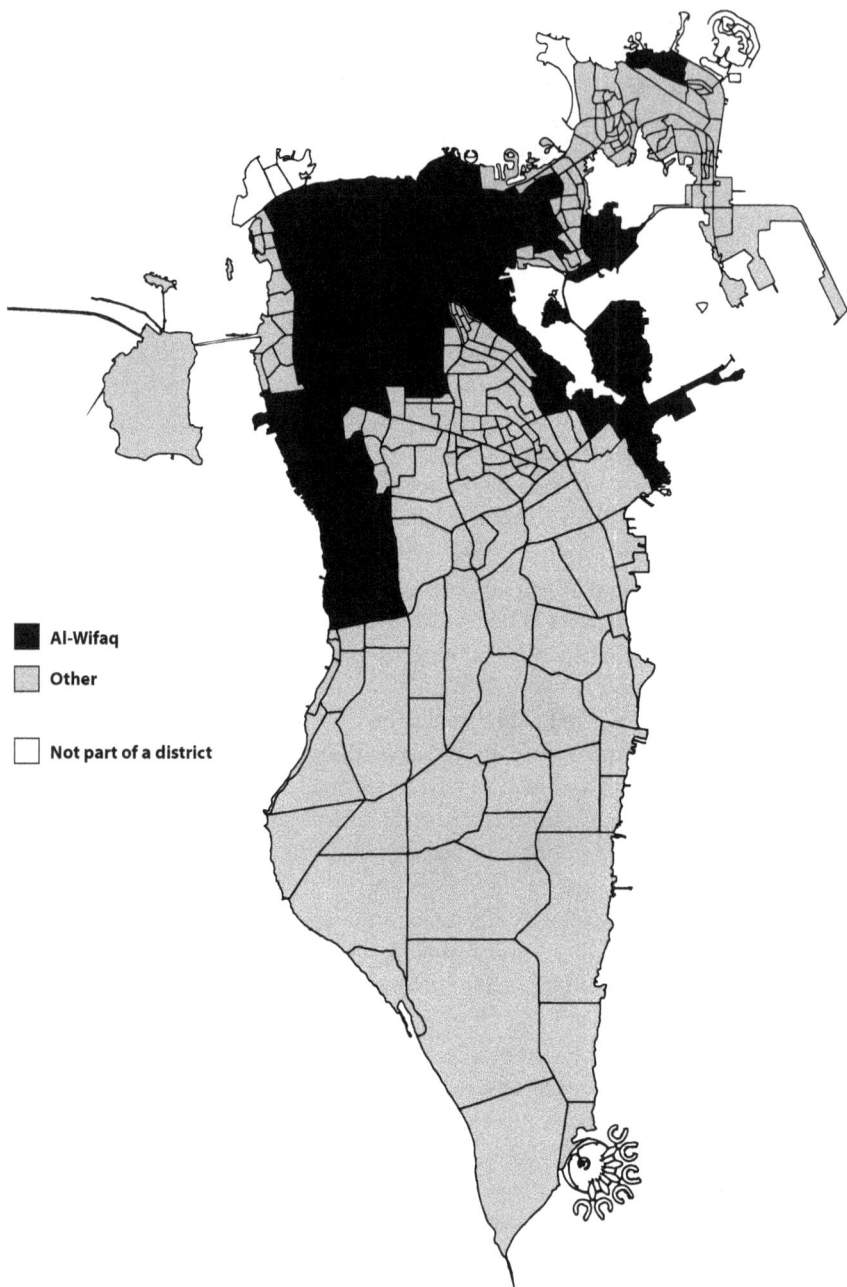

Al-Wifaq

Other

Not part of a district

Map 4.3. Map of Bahraini electoral districts, by 2010 winner.

description of "apartheid-like," nonetheless remains glaring. Of 187 total block numbers visited in the survey, 119, or 64 percent, are represented by more than one interviewee. Of these, only 30, or 25 percent, were not exclusive either to Sunnis or Shi'a, and of these mixed blocks a combined one-half were located either in Isa Town (11 cases) or Hamad Town (6). Away from these two urban developments, there were found just 13 other mixed blocks across the remainder of the island, amounting to 11 percent of the 119 multiple-respondent districts. And even these proportions are likely far too conservative, as included among the 68 single-respondent blocks are doubtless many more Sunni- and Shi'i-exclusive areas. No less than 15 of these 68, for example, correspond to Shi'a villages in Sitra and along al-Budayyi' Road, where one would be surprised to find more than a handful of Sunnis. While thus ameliorated somewhat by the construction of new urban housing settlements like Isa Town and the ever-expanding Hamad Town, Bahrain's systematic sectarian self-segregation[26] shows little indication of having lessened qualitatively almost since the time of the Al Khalifa conquest.

How far this geographical polarization reflects an underlying sectarian-based political divide among respondents, and to what extent the latter supersedes purely economic concerns in shaping the political opinions and actions of ordinary Bahrainis—of our ostensibly-quiescent Gulf rentier citizens—we are finally prepared to learn. Thus we are following the practical and methodological preface to my 2009 Bahrain mass survey, which has addressed the two major questions likely to be raised about the present study. The first is whether the choice of Bahrain as a subject perhaps limits the wider applicability of its findings owing to the country's unique historical and sociopolitical circumstances; that is to say, whether the case of Bahrain perhaps lacks external validity in a way that renders the present inquiry one fundamentally about the sectarian politics of Bahrain rather than one about the efficacy of the rentier state framework in explaining political outcomes in the Arab Gulf.

To this objection our response was twofold. In the first place, there exists already anecdotal evidence in support of a general link between sectarian diversity and domestic political stability among the six GCC states. Moreover, and more fundamentally, since sectarian political tensions from the Gulf stem from the same endogenous and exogenous forces—the unequal distribution of rentier benefits, deliberate state strategies of divide and rule, and a widening geopolitical struggle for influence between Sunni- and Shi'i-dominanted states and their great power patrons—there exists a compelling theoretical argument to explain why the case of Bahrain should be generalizable in degrees to the wider Gulf.

A second methodological objection concerned the lack of additional country observations to which to compare the Bahraini case in a more rigorous cross-country analysis. On this point we conceded that it would have been preferable to have replicated the Bahrain study in one or all of the remaining Gulf nations, the more demographically homogenous countries being particularly attractive targets from a theoretical standpoint. Yet, we said, such an expansion would have further burdened a project that, as it was, often seemed destined for failure even in Bahrain. The many practical difficulties plaguing the execution of the mass survey we need not revisit; it is enough to mention a lack of social anonymity and trust, physical and social isolation of sectarian enclaves, and persistent unrest.

Yet notwithstanding such difficulties, we concluded, there is reason for sanguinity as regards this Bahrain inquiry. For, even if the larger aim of understanding the influence of group division on state stability at the regional level is complicated at present by a lack of comparable data, still there is much to be gleaned from a single case. This is because the more fundamental question remains answerable, and answerable for the first time only, namely whether the individual-level causal story of the rentier state framework, these theoretical assumptions about what motivates the political behavior of ordinary rentier citizens like those of the Arab Gulf, finds substantiation in the actions and opinions of actual Gulf Arabs. We proceed now in this direction.

5   Rentier Theory and Rentier Reality

More than simply offer empirical evidence of a general sectarian political disagreement in Bahrain, the present chapter seeks to evaluate the specific theoretical arguments elaborated thus far in explanation of the larger case of Bahrain—the case of the failed rentier state, unable to buy its way to political quiet. It aims to understand both *how* the state distributes economic benefits at the individual level as well as the political *payoff* of these allocations from the standpoint of the regime. What factors make Bahrainis more likely to receive the rentier benefits of citizenship, that is to say, and to what extent do economic circumstances even determine citizens' political attitudes and behaviors in the first place? In answering these questions, the analysis to follow also serves the greater purpose of revealing how far longstanding assumptions about individual political behavior in the Arab Gulf states accord with the reported views and actions of real-life Gulf Arab citizens.

Evidence of classic rentier dynamics operating in Bahrain would be straightforward to observe in the mass survey data. If economic benefits function mainly as a mechanism of co-optation of would-be political opponents, rather than as a reward for current political supporters, then one should expect to find benefits distributed more or less equally among citizens. Insofar as one should expect any inequality in distribution, indeed, one should expect to see resources disproportionately expended upon those groups and individuals viewed as having a greater involvement in, or latent potential for, opposition, since it is these whom the state perceives as being in foremost need of co-optation. Similarly, if it is true that in rentier societies citizens' political orientations are determined above all by material satisfaction, then such a relationship would be simple to observe in the survey data. In that case, any outward sectarian or other group-based discrepancy in political attitudes or behavior would constitute a mere epiphenomenon masking underlying socioeconomic differences and would disappear when the latter were taken into account.

Alas, such classic rentier dynamics are not in evidence in the Bahrain mass survey data, whether on the supply or demand side. Not only are private and public goods not distributed among Bahrainis in a politically agnostic manner, but also citizens' normative attitudes toward the state, as well as the political actions they take in support of or against it, are shaped by a quite wider set of individual-level variables. Regarding the former question, for instance, consider the results summarized in Table 5.1, which reports Bahrainis' evaluations of various public services. Asked to rate the ease of accomplishing a variety of basic tasks requiring government cooperation—acquiring official documents, such as a passport or birth certificate; enrolling a child in a public school; receiving medical treatment at a nearby state facility; receiving assistance from the police when needed; and accessing individuals and institutions to resolve perceived rights violations—in each case the data reveal a divergence in evaluation between Sunni and Shi'i citizens, with the latter reporting a higher difficulty of accomplishment. For some tasks, such as school enrollment and to a lesser extent acquiring official documents, this empirical discrepancy corresponds to a relatively minor substantive difference, yet for others it is more meaningful. Almost twice as many Shi'a than Sunnis say it is "difficult" or "very difficult" to receive medical care at a nearby government facility, for instance, while nearly half of Shi'a respondents report a similar difficulty in receiving police assistance, compared to just 21 percent of Sunnis. Finally, only a combined 17 percent of Shi'a deem it "very easy" or "easy" to resolve rights violations, compared to a third of Sunnis.

Another public good the survey data reveal as being distributed unequally between Sunna and Shi'a in Bahrain is public security—and this two years before the 2011 uprising. Not surprisingly, perhaps, given the conflicting impressions of police assistance seen in the preceding chapter, in early 2009 Bahraini Shi'a perceived their neighborhoods as being far less safe than did Sunnis, as reported in Table 5.2. Not only this, but Shi'a also were far more likely than their Sunni co-nationals to report a decline rather than improvement in safety over the recent past, with 43 percent of Shi'i respondents—more than three times the proportion of Sunnis—describing their neighborhood as "less safe" than it was three to five years ago.

In short, then, these quantitative results would seem to accord well with the anecdotal impressions of the disparity separating Sunni- and Shi'a-dominated regions of Bahrain noted already in chapter 4 and remarked regularly by observers since the days of the feudal estate system. Two hundred years after the consolidation of Al Khalifa power in Bahrain, the two communities remain physically segregated outside of a handful of urban centers, and the quality of infrastructure and public services is, on the whole, manifestly inferior in the numerous rural villages populated mainly by Shi'a. One will recall the Shi'i respondent who complained of police refusing to enter his village to investigate traffic accidents, requiring residents instead to tow their vehicles to an adjacent main road. Another man in Sitra told how a child drowned when a beach was dredged without warning, the offending company merely erecting a rickety fence in response.

Of course, one might explain the observed difference in perception and evaluation of state services as a difference mainly in the two communities' relative willingness to offer negative opinions rather than as evidence of government neglect. It might be, in other words, that Bahraini Shi'a are simply more disposed

Table 5.1. Evaluation of government services, by sect

Service	Very Easy		Easy		Difficult		Very Difficult		
	Sunni	Shi'i	Sunni	Shi'i	Sunni	Shi'i	Sunni	Shi'i	p-value
Official documents	43%	27%	50%	60%	5%	11%	2%	3%	0.004
Child school enrollment	43%	33%	53%	62%	3%	4%	1%	0%	0.151
Medical care	34%	20%	54%	58%	10%	17%	2%	5%	0.002
Police assistance	26%	13%	53%	39%	16%	30%	5%	18%	0.000
Complaint resolution	7%	2%	26%	15%	46%	35%	21%	48%	0.000

Notes: Final columns report tests of Pearson's F-statistic; sampling weights utilized; $N=310$–427 (some missing/not applicable).

Table 5.2. Evaluation of neighborhood safety, by sect

Measure	Very Safe		Safe		Unsafe		
	Sunni	Shi'i	Sunni	Shi'i	Sunni	Shi'i	p-value
Neighborhood safety	32%	18%	60%	66%	8%	16%	0.001
	More Safe		No Change		Less Safe		
Recent change in safety	40%	21%	43%	36%	17%	43%	0.000

Notes: Final columns report tests of Pearson's F-statistic; sampling weights utilized; $N=416, 431$.

than Sunnis to answer survey questions in a way that directly or indirectly criticizes the state. It is possible that the evaluations of Shi'a are tinged by negative political orientations—not on account of their being Shi'a but for their weaker socioeconomic condition or some other exogenous cause—such that Shi'i respondents tend to give more critical assessments than Sunnis of the same objective phenomena.

This interpretation fits well with the standard explanation of Shi'a activism in Bahrain, namely that the community's frustration is, at bottom, *economic* frustration channeled through religious identities and religious-cum-political institutions such as the clerical establishment, Shi'i political societies and movements, and so on. Thus, for instance, does King Hamad begin his April 2011 apologia in the *Washington Times*, in which he defends the state's violent suppression of mass protests: "On Feb. 14, the winds of change that are sweeping the region hit the

shores of Bahrain. Demands for well-paying jobs, transparency in economic affairs and access to better social services were received with good will." It was only
when the opposition and its legitimate economic grievances were "hijacked by ex-
tremist elements with ties to foreign governments"—Shi'a with an Iranian-backed
sectarian agenda—King Hamad continues, that the state was compelled to act with
the brutal resolve still evident at the time of his writing.[1]

To disentangle the competing explanations of Tables 5.1 and 5.2, accordingly,
is to go a long way toward resolving the larger conceptual debate about the nature of political conflict in Bahrain and the other Arab Gulf states that feature
ascriptive social cleavages. In Bahrain and elsewhere, are certain religious and ethnic constituencies oppositional because they are marginalized politically and economically? Or are they disproportionately excluded from politics and from the
economic benefits of rentierism on account of what rulers perceive as an innate
propensity toward opposition or disloyalty? In the case of Bahrain at least, we can
begin to unravel this endogeneity by delving further into the mass survey data to
investigate, not only the substance, but also the empirical determinants of economic and political outcomes at the individual level.

A "Legitimate Aspiration" for All? The Politics
of Public-Sector Employment in Bahrain

Among the basic tenets of rentier state theory is that no citizen has a greater claim
to public benefits than any other, since the basis of qualification is citizenship itself rather than some personal attribute(s). This principle is encapsulated in Beblawi's assertion visited in chapter 2 that "[e]very citizen—if not self-employed in
business and/or working for a private venture—has a legitimate aspiration to be
a government employee; in most cases this aspiration is fulfilled." Having already
elaborated the conditions giving rise to the other half of these "most cases"—those
cases in which a Bahraini or other Gulf citizen is more likely to be rebuffed than
have his public employment aspiration fulfilled—at last we have the opportunity
to test this counterargument using individual employment data from the Bahrain
mass survey. Beyond the theoretical centrality of the question per se, this analysis of government employment is significant in that it avoids the ambiguity inherent in the preceding treatment of public goods provision, since one's sector of
employment is not a subjective evaluation but a fact that is simply observed.

The aim of this analysis is therefore straightforward: to discover whether sectarian affiliation is a significant predictor of employment status among Bahraini
citizens, in regard to both an individual's sector of work as well as his professional
level. If we find, as anticipated, that one's chances of being a government employee
are reduced materially when one is a Shi'i, and again that being a Shi'i is negatively related to the professional level of one's pubic-sector position, then we will

have evidence that public employment in Bahrain does not operate neatly in the service of popular political pacification as rentier theorists would have it, precisely because it disproportionately excludes those most in need, from the state's perspective, of pacifying.

At first glance, the data from the Bahrain survey would seem to point toward a group-based discrepancy in public-sector employment. Of the 143 Shiʻi respondents employed at the time of surveying, only 55, or 38 percent, reported working in the public sector. By contrast, 52, or 51 percent, of the total 103 working Sunnis reported being state employees.[2] Yet, while a positive indication, these preliminary results are unsatisfactory for two reasons: first and most importantly, the apparent association between sectarian membership and employment sector may mask differences in other relevant individual-level variables such as education level, gender, and so on. Less obvious but even more significant is the fact that this basic model, even if it were to include relevant control variables, is not an accurate representation of the process we seek to explain. Since we observe a respondent's sector of employment only when he or she is employed, the sample of 246 respondents is not a random sample of the Bahraini population but one that is systematically truncated to include only the *employed* Bahraini population. As a result, the apparent between-group difference in public employment may be a function not of group membership per se but rather of a Sunni-Shiʻi discrepancy in employment in general. That is, there may be unobserved variables affecting participation in the workforce—women's participation, education, age, and so on—that are also correlated with sectarian membership and that therefore bias the analysis of the latter's impact on employment sector.

For an accurate test, then, one must adopt an estimation strategy that reflects both stages of this process and so avoids the selection bias implied above: a first stage that models the probability that a respondent, a random individual from among the entire Bahraini population, is employed; and a second that models the probability that this respondent, given that he or she is employed, is employed in the public sector. Fortunately, one has recourse to Heckman's selection model, which carries out exactly this procedure, designed as it was specifically to correct the sample selection bias problem inherent in analyses of workforce participation.[3] The Heckman strategy usually takes the form of a two-equation structural model that employs one or more identifying variables in the selection equation (i.e., the model of workforce participation) to obtain unbiased estimates in the behavioral equation (the model of sector of employment).[4] These identifying variables are such that they influence an individual's chances of being selected (in this case, employed) but *do not* influence the outcome of the behavioral model (being employed in the public versus private sector) except insofar as they do so via their influence on selection itself. When this condition is satisfied, Heckman's method provides unbiased behavioral model estimates no longer affected by unobserved variables.

All this is to say that by using a well-established technique one may offer a robust test of the effect of sectarian membership on government-sector employment in Bahrain. We specify the (behavioral) model as follows:

$$\Pr(SECTOR=1) = B_0 + SECT \cdot B_1 + EDUC \cdot B_2 + SECT \times EDUC \cdot B_3$$
$$+ FEM \cdot B_4 + \lambda \cdot B_5 + \varepsilon,$$

where *sect* is a dichotomous measure coded 0 for Shi'is and 1 for Sunnis, *education* is a categorical variable coded on a four-point scale,[5] *female* is another binary control, and λ represents the inverse Mills ratios estimated in the selection model (see note 4). Finally, the multiplicative interaction term *sect × education* is included to investigate whether the effect of education on the probability of public-sector employment is influenced by one's sectarian affiliation. Are Sunnis and Shi'is of equal educational background, in other words, equally likely to work in the public sector? If it is true that government employment in Bahrain functions as in other rentier states as an economic safety net for those citizens otherwise less likely to find work, then the state sector should be disproportionately filled with those individuals whose attributes render them less employable in private industry, including those with lower educational qualifications and (for cultural reasons) women. On the other hand, such a conception also implies that sectarian affiliation should have no independent effect on the probability one works in the public sector, with any group-based discrepancy vanishing once the effects of education, gender, and other variables are statistically controlled.

What remains, then, is the (selection) model of the determinants of employment proper, which we express as the following:

$$\Pr(WORK=1) = B_0 + SECT \cdot B_1 + FEM \cdot B_2 + EDUC \cdot B_3 + MARRIED \cdot B_4$$
$$+ MARRIED \times FEM \cdot B_5 + AGE \cdot B_6 + AGE^2 \cdot B_7 + \varepsilon,$$

where a respondent's marital status[6] and age[7] serve as identifying variables. These indicators one should expect to be significant determinants of employment status, but *not* of one's sector of employment. About the predicted effects of the independent variables perhaps little needs to be said. The female gender indicator one should expect to be a strong negative predictor of employment particularly among married individuals;[8] education level, coded on a four-point scale, a strong positive predictor irrespective of other attributes; and marital status a strong positive predictor among males. (Few families would consent to a marriage to an unemployed prospective husband.) Finally, the probability of employment should increase with age until a certain point, and thereafter decline as older individuals drop out of the workforce. The indicator of sectarian membership is included here again in the selection model to confirm that its effect on public-sector employment is not an artifact of its effect on employment generally. Group affiliation should remain a significant predictor of public-sector employment even after controlling its effect on employment proper. Indeed, there is no *a priori* reason to believe that sectarian membership should be related to workforce participation itself.

The full results of this and other estimations are reproduced in the appendix. It is sufficient to say here that the observed Sunni-Shi'i discrepancy in public-sector employment persists even after controlling other relevant individual-level variables and correcting the model's specification. Holding constant the influences of age, gender, and education, the predicted probability of being employed in the government sector, conditional on being employed, is an estimated 41.7 percent for a Shi'i respondent, compared to 65.1 percent for a Sunni. This between-group difference is significant at the 0.001 level. In relative terms, a Sunni Bahraini is 56 percent more likely to be employed in the public sector, given that s/he is employed, than a Shi'i of identical gender, age, marital status, and education level—that is to say, a Shi'i of identical employment-relevant attributes. By contrast, sectarian membership has no independent effect on the chances of employment per se, as demonstrated in Figure 5.1.[9]

Notably, however, this Sunni-Shi'i gap in state sector employment is not a universal one but rather exists for a specific class of observations that drive the overall result. Specifically, the gap operates via an increased probability of government employment among low-education Sunnis as compared to low-education Shi'a. The socioeconomic safety net supposed to exist in rentier states in the form of guaranteed employment, in other words, would seem in Bahrain to exist disproportionately for the benefit of the Sunni community. Depicted in Figure 5.2 is the predicted probability of public-sector employment disaggregated by sectarian membership and discrete education level, with bands showing 95 percent confidence intervals. From this illustration one easily perceives where the discrepancy lies: whereas a Shi'i of primary or lower education is estimated to have only a 23 percent conditional probability of being employed in the state sector, Sunnis of the same education level are effectively guaranteed a government position. Though lesser in magnitude, a similar gap exists for citizens with a terminal secondary education, with Shi'is possessing a high school diploma estimated to have just a 36 percent chance of working in the public sector, compared to a predicted 62 percent among Sunni high school graduates. The statistical significance of the Sunni-Shi'i gap in this case is also less robust, but with an associated *p*-value of 0.025 the odds of observing such a difference by random chance are approximately 40 to 1.

By contrast, no such discrepancy exists among individuals of the two highest education levels, who, independent of sectarian affiliation, are no more or less likely to work in the public sector versus the private. Such is perhaps to be expected, as their higher qualifications open a range of employment and professional options not enjoyed by individuals of lesser education. Yet, as the lowest two educational

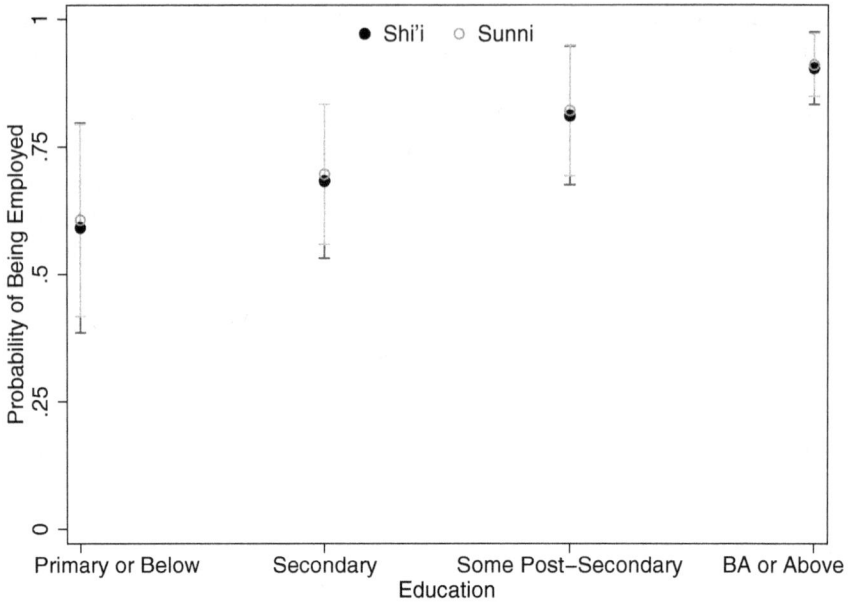

Figure 5.1. Predicted probability of employment, by education and sect.

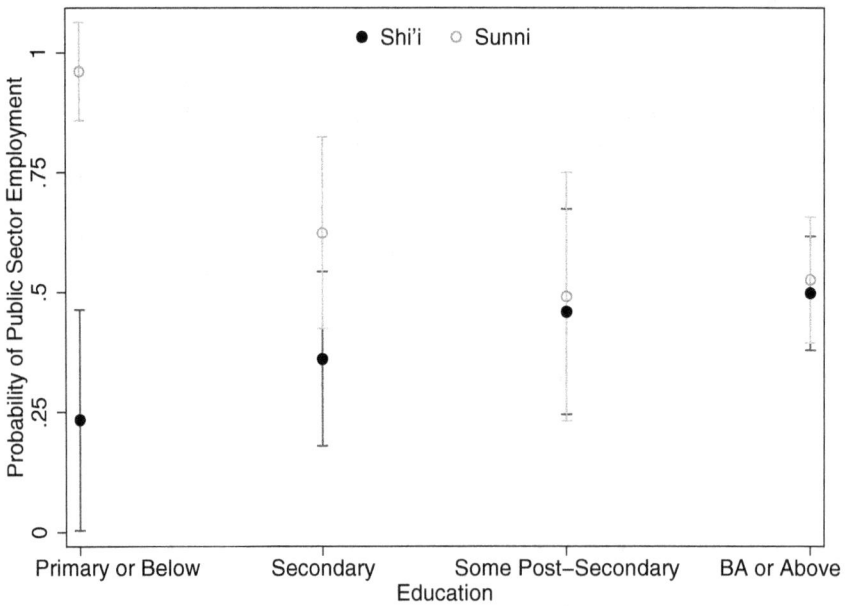

Figure 5.2. Predicted probability of public sector employment, by education and sect.

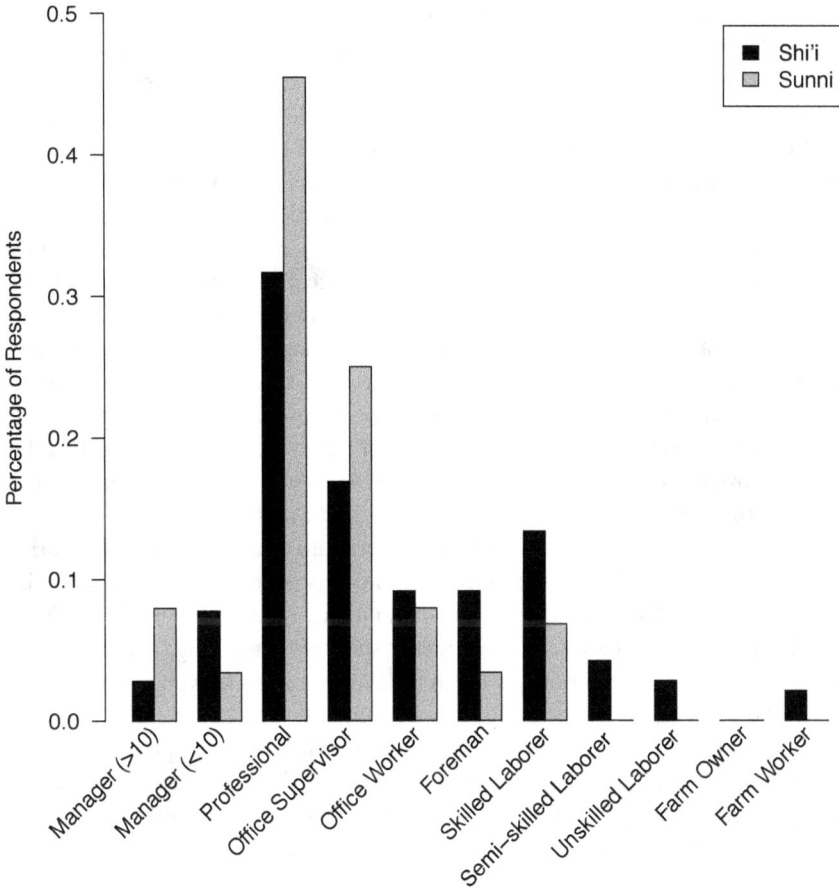

Figure 5.3. Occupational level among employed Bahrainis, by sect.

categories represent 14 percent and 31 percent of all Bahraini respondents, respectively, the disproportionate ease with which less-educated Sunnis are able to secure work in the government sector, and the simultaneous difficulty of this for Shi'a, should not be understated. The rentier state's financial safety net, to say it once more, would appear not to be unfurled universally for all Bahrainis.[10] Not every citizen enjoys an equally "legitimate aspiration to be a government employee."

Furthermore, work comes in myriad types, and even the promise of employment is no guarantee of a job commensurate with one's qualifications. The next step, then, is to consider the determinants not simply of the sector of employment among Bahrainis, but the character and status of that employment. Survey respondents were asked to place themselves on a 13-point professional scale ranging from "an employer/manager of an establishment with 10 or more employees" to "an agricultural worker."[11] Appended to this descending scale were two additional options that elude easy categorization: members of the armed forces and police, and housewives. The latter for our purposes we need not consider, but the question of how accurately to classify military and police personnel is a more difficult one and, as we shall see, one that has a considerable impact on what we conclude about the effect of sectarian membership on occupational level in Bahrain.

This is because, as discussed briefly in chapter 3, not a single Shi'i of all those interviewed reported being employed in the police or military,[12] compared to 12 (or 12.1 percent) of the 99 working Sunnis who gave their occupations, of whom just one was a female.[13] Among employed Sunni males who reported occupational data, 11 of 66 (or 16.7 percent) said they worked for the military or police, compared to 0 among 117 working Shi'i males who reported data. So that, even if we include both sexes, we arrive at an estimate of 1 in every 8¼ Bahraini Sunnis being employed in the state security services. Moreover, when we add the data that respondents provided about their spouses, we find that 5 (or 7.2 percent) of the 69 married Sunnis who reported their spouse's occupation indicated that s/he worked in the military or police. Aggregating the two sets of responses, finally, we discover that these 168 observations correspond to just 131 unique Sunni households in which a respondent and/or a respondent's spouse was working. This means that of the 131 working Sunni households in the Bahrain mass survey, a minimum of 17, or 13.0 percent, are police or military households.

Beyond seeming to vindicate those Bahraini Shi'a who complain of their exclusion from the police and military, the fact that we have here a Sunni-exclusive category comprising 12.1 percent of all Sunni respondents is also of more immediate significance, for where it is placed on the scale of occupations will necessarily have a great statistical influence on any estimated relationship between sectarian membership and professional level in Bahrain. In which job category does this group belong? Surely, its present, concluding position below even farmers and agricultural workers makes little sense. Yet should one deem police and military personnel "professional workers" of category 3 along with teachers and accountants? or "skilled manual workers" of category 8 along with mechanics? Or should the military/police category itself be moved to some other position along the existing scale? Further, given the heterogeneity of ranks within the military and police, should one assume that all respondents are rank-and-file soldiers? commanding officers? or some level in between? Ultimately, such questions point to the safest course of action, which is to omit the category altogether from the quantitative analysis of professional level. If we thus lose a bit of statistical leverage on our question about the relationship between occupational level and group membership in Bahrain, at least we shall avoid making conclusions about it that are unduly influenced by a single category that in any case seems to be out of place, both spatially and theoretically, among the others.

Even with this omission of the military/police sector, however, one finds no lack of Sunni-Shi'i occupational discrepancy in Bahrain. Consider Figure 5.3, which depicts the professional categories reported by employed Bahraini respondents, disaggregated by sectarian affiliation. One sees that, overall in the economy, Sunnis are relatively better represented than Shi'is as managers of large establishments, professionals, and office supervisors. By contrast, Bahraini Shi'a are relatively better represented as managers of small establishments, non-supervisory office workers, foremen and supervisors of manual labor, and skilled manual workers. Moreover, Shi'i respondents alone fill the categories of semi-skilled and unskilled manual workers, as well as agricultural workers. No respondent reported owning his own farm. Certainly, this would appear to be strong evidence that Bahrain's executive, supervisory, and professional classes are disproportionately occupied by the nation's Sunnis.

But, as in the case of employment sector, these relative proportions are insufficient to evidence a general relationship between sectarian membership and occupational level. One encounters again the same two problems: the seeming pattern of Figure 5.3 may disappear or change when one adds relevant control variables; and the sample is truncated to include only those Bahrainis who are employed. Furthermore, the relationship of primary theoretical interest is not this between sectarian membership and occupational level generally, but that between group affiliation and occupational level among public-sector workers in particular. One must therefore use an estimation strategy analogous to that adopted previously: one that models the selection process inherent in the occupational data, but one that also can be restricted by sector of employment. This two-part model utilizes the same selection equation for employment seen already, whereas the behavioral equation becomes:

$$JOB = B_0 + SECT \cdot B_1 + EDUC \cdot B_2 + SECT \times EDUC \cdot B_3 + FEM \cdot B_4 + AGE \cdot B_5$$
$$+ \lambda \cdot B_6 + \varepsilon, \text{ if } PUBSECTOR = 1.$$

Compared to the foregoing analysis, then, the model setup here is little changed: only the behavioral equation is estimated now by ordinary standard least-squares given the continuous measure of occupational level. The explanatory variables remain the same, with the exception of one additional indicator, age, which one might assume to be positively associated with occupational level in line with the idea of seniority. Also as before, included in the behavioral equation is a final regressor λ, which is a vector of inverse Mills ratios computed from the selection model (serving, again, as a control for sample selection bias; see note 4). Lastly, in order that individual cases do not unduly influence the results, the outlying observations in categories 1 and 11—see Figure 5.4—are omitted from the analysis.[14]

The results of this estimation are strikingly familiar. Just as in the case of Bahrainis' sector of employment, here the professional level of workers in the public

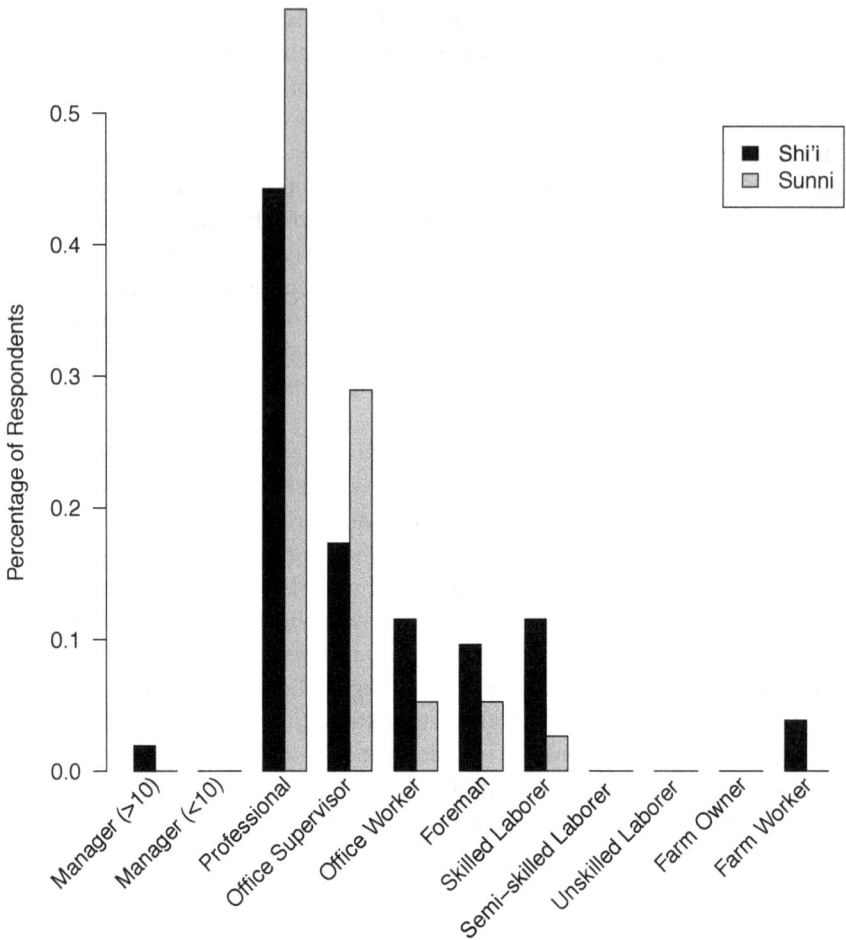

Figure 5.4. Occupational level among Bahrainis in the public sector, by sect.

sector is shown to depend crucially upon a specific combination of sectarian affiliation and educational level. This interaction is depicted visually in Figure 5.5,
which gives the predicted professional level of Sunnis and Shi'is according to terminal education. Here again, Sunni citizens tend to enjoy an advantage over their
Shi'i co-nationals with regard to professional status, but this discrepancy declines
with education to the point that, among college graduates, there is no statistical
Sunni-Shi'i difference in job level at all. By contrast, among individuals with a
primary terminal education or less, Sunnis enjoy an estimated 1.9-category advantage in occupational level even after accounting for the independent effects of

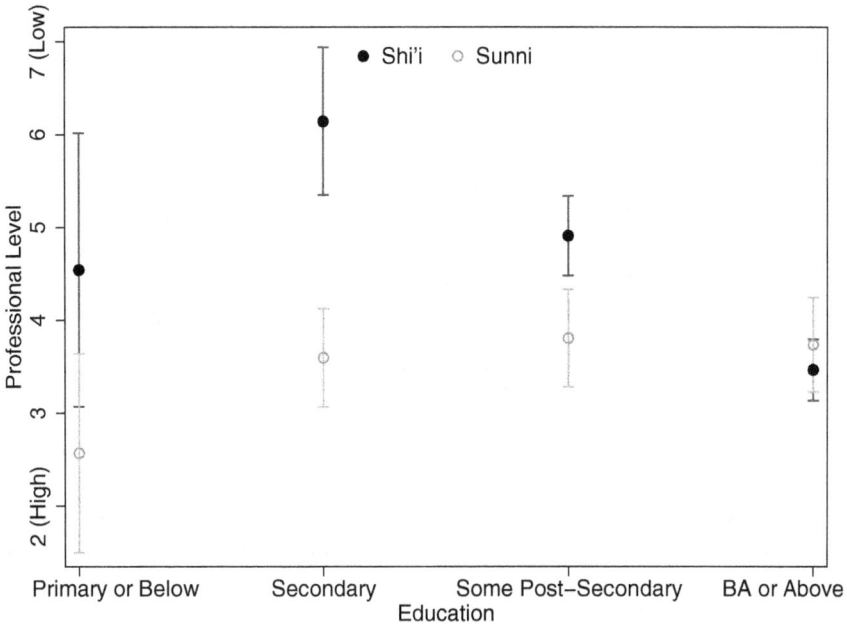

Figure 5.5. Predicted public-sector professional level, by education and sect.

gender, age, and a respondent's baseline probability of being employed. At an estimated 2.5 categories, this advantage is even greater among citizens with a terminal secondary education. It is reduced to an estimated 1.1 among those with some post-secondary schooling, and, as indicated already, the between-group difference disappears entirely among college graduates.

Yet, even more notable than this Sunni-Shiʿi gap per se is its underlying cause: an obvious disconnect between educational qualifications and professional status among Bahraini Sunnis. Excepting for now the lowest educational category, whose relationship with job status is more ambiguous owing to few (3) observations, the remainder of Figure 5.5 reveals two very different patterns applying, respectively, to Bahrain's two confessional communities. Whereas one observes among Shiʿa the expected link whereby those of higher educational achievement tend to achieve in turn higher occupational status, among Sunnis working in the state sector there is no statistical connection between one's educational and job levels. The predicted occupational category of a Sunni with a college diploma is no different from that of a mere high school graduate. By contrast, among Shiʿa employed in the public sector a college diploma dramatically improves one's expected professional level, from an estimated 6.1 among secondary school graduates to around 3.5 among individuals with a bachelor's degree or higher. In real terms, this estimated statistical divergence among Shi'a means the difference between one's working as a labor foreman and somewhere between a professional and office supervisor. Sunnis in the government sector, on the other hand, can expect to fill one of the latter types of positions irrespective of educational background and other objective attributes such as age and gender.

Thus there emerges in this study of employment in Bahrain a persistent pattern: whether with regard to one's sector of work or to the substance of that work, Sunni citizens seem to enjoy a considerable advantage over their Shi'a compatriots. It is not simply that Sunnis are overrepresented in the state sector and disproportionately fill professional and supervisory roles in public entities, but that the entire selection process for employment itself seems to operate in a fundamentally different manner according to a citizen's sectarian affiliation. The standard conception of public employment in rentier states, as a sanctuary for those less suited for the private-economy workforce, would seem to obtain primarily among Sunnis. Not only are lesser-educated individuals from among this community far more likely than similarly qualified Shi'a to work in the government sector, but moreover their relative lack of educational credentials does not serve as a professional hindrance. Shi'a are not extended the same courtesy: only individuals with some college education are represented in the state sector in equal proportion with Sunnis of equivalent employment-relevant attributes; and only university graduates can expect an occupational status on par with Sunnis of similar credentials. For the remaining 35–40 percent of working-age Bahraini Shi'a with a secondary terminal education or less, inequality in both the opportunity and nature of public-sector employment is more than an anecdotal impression colored by political orientation—it is an empirical reality.[15]

Beyond its intrinsic significance, this finding also affords some empirical leverage for answering the larger question with which this discussion began: namely, do the more negative impressions of state benefits among Bahraini Shi'a reflect a genuine gap in public goods provision between the two communities, or simply a more negative orientation among Shi'a in general? That is, are Shi'a citizens less satisfied with government benefits than Sunnis because they are, objectively speaking, more poorly served, or because they allow their more negative *political* orientations toward the state to color their evaluations of it in other areas? Which came first: economic or political discontent?

To all appearances, the observed patterns of public-sector employment in Bahrain would seem to support the former interpretation. Despite having no more difficulty than Sunnis securing employment overall, Shi'a citizens—less-educated Shi'a specifically—face considerable barriers with respect to the government sector in particular, in terms of both employment and professional status. So, either Shi'a of lower education systematically (but for reasons unclear) spurn the public sector in favor of the private,[16] or else they experience some form of discrimination in government hiring. Further, that Shi'a experience difficulty in taking advantage of this the greatest of all material benefits supposed to be provided by distributive states, makes very plausible the idea that they would find themselves at a similar disadvantage with respect to other public goods and services meant in principle to be distributed equally among citizens. If it is true that Sunnis enjoy a demonstrable privilege in this area, why not in others as well?

Whatever the case, it is clear that the classic rentier understanding of government employment and other benefits afforded citizens, the notion that in allocative regimes the public sector serves as a sort of parallel job market that, if less demanding of its aspirants, remains fundamentally economics-driven and politically agnostic—it is clear that this conception cannot tell the whole story. Instead, here as elsewhere we find evidence that the ostensive material underpinning of the resource-based economy, the cold economic bargain said to exist between Gulf rulers and their clients-cum-citizens, is a rather more pragmatic partnership, colored and adulterated by other, competing considerations. In Bahrain at least, the smooth translation of resource wealth into tangible material benefits for *all* citizens seems not to operate so smoothly after all.

Understanding Individual Political Behavior in Bahrain

The state's over-provision of benefits to Sunnis as compared to Shi'a suggests an important implication about the link between material and political satisfaction in Bahrain. Specifically, insofar as one ascribes to this advantage a *political* rather than fundamentally sectarian cause, then it follows that from a political standpoint Sunnis should be relatively more sensitive to economic concerns than Shi'a. If Sunnis receive a greater share of scarce national resources as a reward for their (presumed) superior political loyalty, then the ostensive rentier bargain assumed to operate in Bahrain in fact applies primarily—one might even say only—to Sunnis. As such, one should find that the political attitudes and behavior of Sunnis are relatively more responsive to economic conditions at the individual level than are Bahraini Shi'a, as Sunnis expect material compensation for their political support. Shi'a citizens, by contrast, are not truly party to the tacit benefits-for-allegiance agreement in the first place, so their expectations of material reward are dampened. Shi'a therefore have far fewer incentives to exhibit politically desirable opinions and behavior, these being effectively decoupled from personal economic gain.

Combined with the theoretical discussion of chapters 1 through 3, this insight brings to four the number of testable hypotheses about the individual-level determinants of political attitudes and behavior in Bahrain suitable for exploration in the mass survey data. These are:

> H1. Confessional affiliation is the dominant factor shaping Bahrainis' political opinions and behavior, with Shi'a citizens possessing more negative orientations.

H2. Heightened confessional identification augments positive orientations toward the state among Sunnis and negative orientations among Shi'a.

H3. Heightened confessional identification augments political activity among members of both communities.

H4. The effect on political orientations of personal economic circumstances operates more strongly and systematically among Sunnis than Shi'is.

These four propositions represent a basic *a priori* critique of the extant rentier paradigm as it applies to Bahrain and other ascriptively divided societies. Rather than an enticement to win over opponents, economic benefits are distributed as a reward for those identified, mainly on the basis of observable descent-based characteristics, as current supporters. The latter have then a strong personal stake in the preservation of the political status quo. The state need not buy off all citizens, accordingly, and indeed in Bahrain and most other Gulf countries lacks the resources to do so in any case. There simply is no universal rentier bargain tying all citizens to the regime *qua* material benefactor. Members of the ascriptively defined in-group can reasonably expect to benefit from maintaining at least the outward appearance of loyalty; others, irrespective of their observable views and actions, have no such expectation, and thus no incentive beyond fear of physical reprisal to be or appear politically supportive.

To keep the analysis as parsimonious as possible, these predictions will be tested simultaneously within a single standard model varying only in the choice of dependent variable. As the first hypothesis simply implies a between-group difference in a given response, it can be assessed through the inclusion of the same dichotomous indicator of confessional affiliation that was used in the foregoing analysis of public benefits. Hypotheses 2 and 3, on the other hand, together make a relatively more complicated argument: that Bahrainis' political orientations are influenced not only by confessional membership, but also by the strength of one's identification with that group; and that, moreover, the substance of the latter effect differs between Sunnis and Shi'is. That is, the political orientations of Bahrainis are determined in part by an interaction between the effects of two variables: sectarian membership itself, along with the personal salience of that group identity (operationalized here as religiosity[17]). This claim can be evaluated with the inclusion of a multiplicative interaction term, while an analogous interaction term involving a measure of household economy allows a straightforward test of the final hypothesis.

For the measure of respondent religiosity one is faced with several alternatives. One commonly used measure is the frequency with which one performs religious rituals such as prayer, reading of the Qur'an, or mosque attendance. However, the difficulty with such measures is that, due to the practical differences between Shi'i and Sunni Islam, the substantive meaning of, say, "mosque attendance," is likely to differ across the two groups. For a Bahraini Shi'i, for example, this may be taken to include religious services at a *ma'tam* or for a religious commemoration. Likewise, Shi'is, who in Bahrain as elsewhere seek edification in "Husaini" literature, may report reading the Qur'an less than Sunnis of equal religiosity, rendering that measure inconsistent. Moreover, all but overtly secular respondents will probably tend to exaggerate the frequency of their observance lest they appear irreligious, an incentive operating upon Sunna and Shi'a alike.

For the same reason one must be careful too in employing the other standard survey item meant to gauge religiosity, namely a direct question. In this case, the standard Arab Barometer instrument asks, "In general, would you describe yourself as a person who is religious [*mutadayyin*] or non-religious [*ghayr mutadayyin*]?" In the first place, once again apart from deliberately secular individuals, a typical Arab Muslim of any religiosity is unlikely to wish to describe himself in this fashion as "non-religious" and so will tend to default to "religious." In the second place, the most commonly used word for "religious" here carries overtones in Arabic of someone who is not simply religiously observant but unusually devout and perhaps even fanatical in his or her faith. It is no wonder, then, that of the 389 Bahrainis who responded to this question, 57 (or 15 percent) were unsatisfied with both choices and so asked the interviewer to record an alternative answer: "*mu'tadil*," or "moderate"; and one assumes that many others who were less concerned for the accuracy of their response simply picked one or the other option.

An alternative method of measuring religiosity, then, is to infer it indirectly from one's response to one or more indirect questions. If the resulting indicator enjoys perhaps less outward validity than one based on a direct query, still it is less affected by social desirability bias and, in reference to the first class of indicators discussed above, is also consistent across the two religious subgroups. The religiosity measure we will adopt for our analysis is of this final type, constructed on the basis of the question: "Which of the following factors [would] constitute the most serious impediment to your acceptance of the marriage of your son, daughter, sister, or brother?": "lack of prayer [on the part of the betrothed], lack of fasting, the social status of the [betrothed's] family, poverty, lack of education, lack of employment, or something else." Our religiosity variable is coded 1 when a respondent identifies either or both of the first two factors as being the most important, or adds an "other" category response that invokes religion. Among the latter instance are included, for example, responses that the most important attribute of the betrothed is reputable "ethics" or "morals" (*al-akhlāq*), "religiousness" (*al-tadayyun*), or simply "belief in God." Several individuals even state explicitly that the most important thing is that s/he be a Sunni or Shi'i. That the "other" category would in this way be filled almost entirely with religious stipulations, including with confessional stipulations, would seem to speak well for the validity of this measure as a proxy for an individual's level of religiosity.[18]

The other independent variable of primary interest is a respondent's personal economy. As its name implies, this is not a measure of income or wealth per se, but instead one that captures a subjective evaluation of overall economic well-being. This remains faithful to the rentier causal story, whose expectations about individual political behavior are based not on finite measures of citizens' income or prosperity but according to overall economic *satisfaction*. In the absence of a more direct question about the latter, this general measure—with operative categories "very good," "good," and "bad"—will serve as a satisfactory substitute.[19]

Control variables include the standard demographic indicators of age, gender, and education level utilized already in the study of employment. More notable, however, are two additional measures that aim to account for possible survey error due to the effects of social desirability. A first, alluded to in the discussion of survey procedure in chapter 4, indicates whether a respondent happened to be interviewed by a fieldworker of the opposite confessional group. This variable allows us to evaluate the possibility that respondents' answers, especially answers to sensitive sociopolitical questions, are likely to be influenced by the (outwardly observable) sectarian affiliation of their interviewer. The resulting interviewer effect also offers one straightforward metric by which to gauge the extent of communal mistrust in Bahrain.[20] Another control measures respondent anxiety more generally, given the charged political context surrounding the survey fieldwork along with Bahrainis' unfamiliarity with public opinion surveys. It counts the number of times a respondent refused to answer a select group of the survey's most sensitive political questions. The resulting sum is then divided by the total number of "sensitive" questions, generating a 0 to 1 measure of respondent refusal. Finally, in case of possible Sunni-Shi'i difference in the effects of these control variables, five corresponding interactions with sectarian membership are also included in the regression.

The result is the following model of individual political behavior in Bahrain:

$$RESPONSE = B_0 + SECT \cdot B_1 + DIFFSECT \cdot B_2 + DIFFSECT \times SECT \cdot B_3 \\ + ECON \cdot B_4 + ECON \times SECT \cdot B_5 + RELIG \cdot B_6 + RELIG \\ \times SECT \cdot B_7 + FEMALE \cdot B_8 + FEMALE \times SECT \cdot B_9 \\ + AGE \cdot B_{10} + AGE \times SECT \cdot B_{11} + EDUC \cdot B_{12} + EDUC \\ \times SECT \cdot B_{13} + REFUSE \cdot B_{14} + REFUSE \times SECT \cdot B_{15} + \varepsilon,$$

where *response* is a survey item of interest. For continuous dependent variables, the model is estimated by ordinary least-squares; for ordinal, by ordered probit; and for dichotomous measures, by probit.

The Sunni Premium

The question of how far sectarian affiliation can explain mass political behavior, whether in Bahrain or elsewhere, is a highly charged one. Far from a purely academic exercise, such an inquiry dives headlong into territory that is uncomfortable both from a local social and political standpoint and to a large extent from an epistemological one. To attempt to understand complex political views and actions with such a crude explanator as sectarian membership, it is said, is to risk falling into the trap of facile, outdated tropes about the sources of conflict in Arab and other developing societies: primordial hatreds, feuds, and rivalries more at home in the Middle Ages than the twenty-first century.[21] Perhaps worse, to do so also inevitably reinforces the very narratives seized upon and cultivated by self-interested rulers, who justify autocracy with the need to serve as hegemonic mediator between otherwise warring factions of society.

Yet, for now at least, the origins of confessional religious division in Bahrain is an issue that can be separated from the attendant empirical question, which asks simply: is there or is there not a qualitative difference in political views and behavior, all else equal, between Sunni and Shi'i citizens in Bahrain? And to this more straightforward question, the mass survey data answer unambiguously in the affirmative. Confessional affiliation is by far the most powerful and consistent predictor of Bahrainis' views and behavior across the full spectrum of political life. From perceptions of state corruption to electoral participation to general attitudes about the role of religion in politics, Sunni and Shi'i Bahrainis exhibit fundamentally divergent opinions and actions.

Consider, for instance, what is perhaps the survey's most general question gauging popular orientations toward the state. Respondents were asked to rate on a ten-point scale their overall satisfaction with government performance. The results, depicted in Figure 5.6, reveal an almost total sectarian-based polarization. Whereas some 90 percent of Sunnis report being more satisfied than unsatisfied (that is, give the state a score of 5 or above), an almost equal proportion (82 percent) of Shi'a express exactly the reverse opinion, with a full 36 percent reporting a satisfaction level of just 1. Thus, at the same time that a full third of Bahraini Shi'a assign the government the lowest possible grade of satisfaction—with a few memorable respondents going even further to offer such responses as "0," "–1," "*taḥt al-arḍ*" ("below the ground"), and, most humorous of all, "Is there any choice lower than 1?"—at the same time, less than one in ten Sunnis reports anything more negative than a mere neutral evaluation of government performance. Whatever the influence of intervening variables such as economic well-being, then, it is apparent already that it will be difficult to match the immense impact of confessional group membership, which for this question accounts for some 40 percent of total variation in response.[22]

This same Sunni premium—substantially more pro-government views and behavior independent of material or demographic conditions—one witnesses across the board in the Bahrain survey data. Table 5.3 presents this Sunni-Shi'i divergence in response for only a sample of pertinent political indicators, estimated by the standard multivariate model specified already above. In all but one instance,

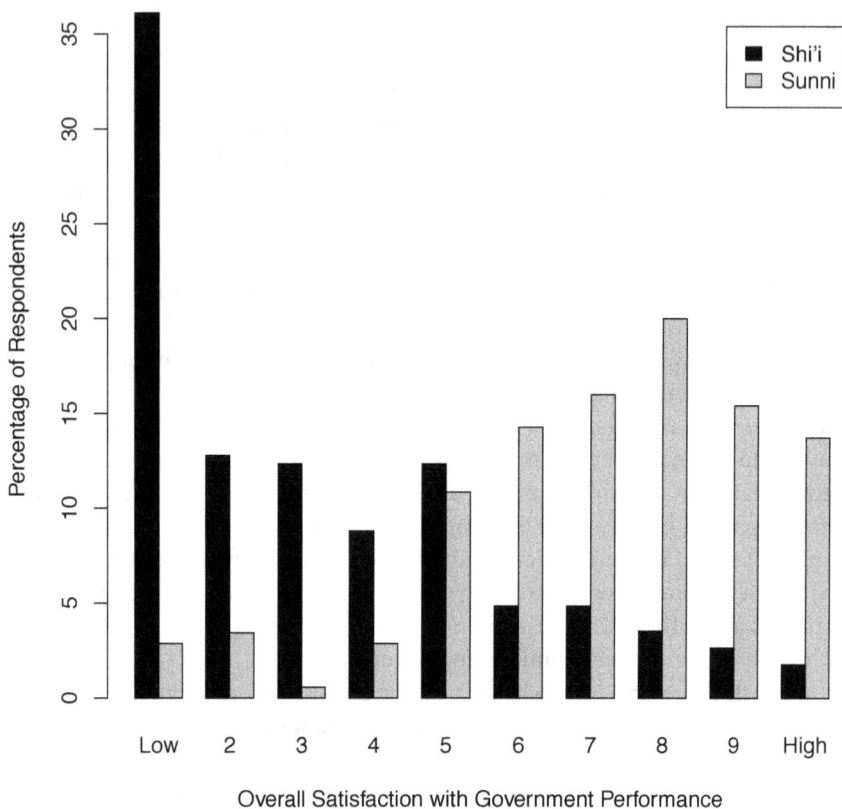

Figure 5.6. Overall satisfaction with government performance, by sect.

the predicted response of a Sunni of average economic and demographic charac-teristics is substantially more favorable than that of an otherwise identical Shiʻi. Sunnis are more sanguine about the influence of government policy on their daily lives, about the country's overall political situation, about the extent to which the country is being ruled democratically, and about the state's efforts to combat cor-ruption. They demonstrate more deference to government decision making, and they are more likely to engage in encouraged political activities such as voting while refraining from unsanctioned behavior like demonstrations. Sunnis are more trusting of basic state institutions such as the judiciary, parliament, police, and the powerful and feared prime minister. In a separate question not reported in Table 5.3 for the sake of space, less than two in three (59 percent) Shiʻis said they would turn to the courts if they were wronged by another person, compared to 83 percent of Sunnis. Still other Shiʻa respondents qualified their answers, saying

Table 5.3. The Sunni-Shi'i divergence in Bahrainis' political views and behavior

	Predicted Response[a]		Sunni Premium[b]
	Shi'i	Sunni	
Evaluation of			
"Overall government performance"	3.2	7.2	125%
(1 worst, 10 best)			
Bahrain's level of democracy	2.7	6.1	126%
(0 lowest, 10 highest)			
"Influence of government policy on daily life"	2.3	3.4	48%
(1 most negative, 5 most positive)			
Freeness and fairness of most recent election	2.3	3.4	48%
(1 neither free nor fair, 4 totally free and fair)			
"Government's management of the economy"	2.5	3.1	24%
(1 worst, 4 best)			
Bahrain's "overall political situation"	1.9	2.8	47%
Trust in			
The judiciary	2.0	3.3	65%
(1 least, 4 most)			
Parliament (elected lower house)	2.0	2.6	30%
Political societies	2.6	2.1	−19%
The police	1.9	3.2	68%
The prime ministership	1.7	3.5	106%
Agreement that			
"Citizens should always support government decisions, even if they disagree with them."	1.8	2.8	56%
(1 strongly disagree, 4 strongly agree)			
The state is working to crack down on corruption	2.0	3.2	60%
Participation in			
A demonstration in the past 3 years	49%	17%	−65%
The most recent parliamentary election	59%	78%	32%

[a] Post-regression linear predictions calculated using margins in Stata 12; evaluated at overall means of all other regressors. Predictions for "Participation in" category represent probit probabilities.
[b] All between-group differences are significant at the $p < 0.000$ level. Sampling weights utilized.

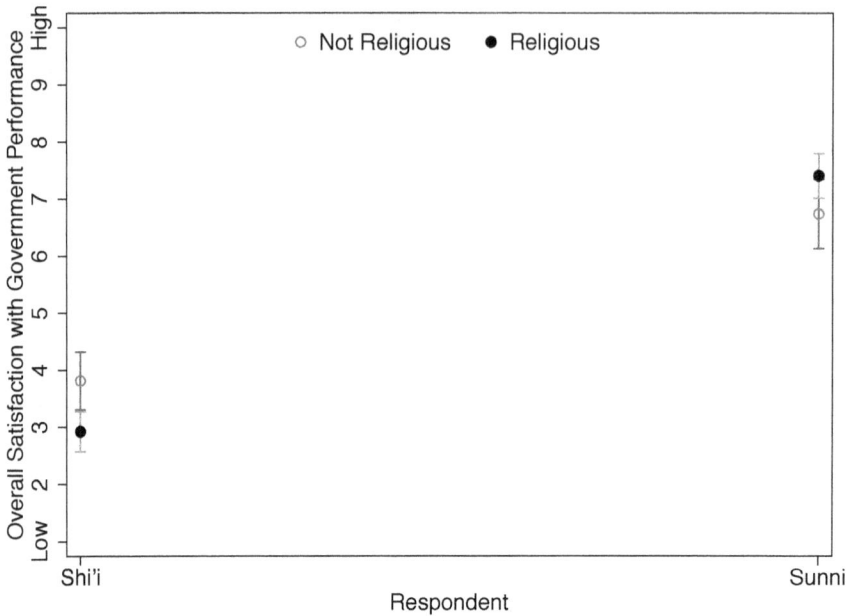

Figure 5.7. The divisive impact of personal religiosity on political attitudes.

that they would go but, variously, "I don't trust them," "they won't do anything," "it's a very difficult process," and "they are corrupt and not independent."

Notably, the one case in which Sunnis exhibit more negative orientations than Shi'a is that of political societies, and this despite their being more positively disposed toward the parliament itself. Shi'a, that is, tended at least at the time of surveying to have more trust in the political representatives of their community but less trust in the institutions they must navigate in order to affect policy. Sunnis exhibited the reverse apprehension, tending to support the institutional framework for representation but not those groups that have formed ostensibly on their behalf. Such a dichotomy accords with the differing organizational choices of the Sunnis and Shi'is who mobilized as part of the 2011 uprising. Whereas the latter largely acted within the existing constellation of opposition actors (with the exception of the youth-oriented February 14th Movement and its offshoots), Sunnis sidestepped the established religious-based societies and their leaders to form grassroots coalitions of a more populist or Sunni nationalist character.[23]

This baseline Sunni-Shi'i political divide is further augmented or assuaged, the survey data reveal, according to the personal salience of Bahrainis' sectarian identities. Among Sunnis, higher levels of religiosity correspond to more favorable views of the state compared to individuals who exhibit less concern for religion.

Among Shi'a, on the other hand, the effect works in the opposite direction, with measures of religiosity associated with more anti-government views. At the same time that stronger in-group identification pushes Shi'a toward more adversarial political orientations, then, it marshals Sunnis further to the regime's defense. This magnifying effect of personal religiousness is depicted in Figure 5.7, which revisits the question of satisfaction with overall government performance. One sees that the predicted response of religious Shi'is, controlling for other factors like age, education, gender, economic situation, and so on, is around one category lower on the 10-point scale of satisfaction than that of non-religious Shi'is. For religious Sunnis, on the other hand, the change is in the opposition direction, boosting satisfaction by around two-thirds of a category, all else equal.[24] The aggregate effect of religiosity, significant at the 0.004 level, is to separate the already disparate evaluations of Sunna and Shi'a by an additional one and a half categories, or 16 percent of the total range of valid responses.

This same substantive impact—buoyed pro-government views among Sunnis and more negative ones among Shi'a—recurs in the survey data across a wide range of political indicators. Although it cannot compete in magnitude or statistical significance with the overwhelming influence of confessional affiliation, nonetheless it does further bolster the notion that Bahrainis' political identification is tied inextricably to sectarian identification. Personal religiousness is not a robust predictor of political attitudes because more pious individuals form a specific political constituency in their own right; not because citizens concerned over religious matters make specific demands upon the state compared to their more secular counterparts. In that case, the effect of religiosity should operate in the same direction among Sunnis and Shi'is alike. Instead, the results show that heightened religiosity represents a stronger identification not as a religious person in general, but *qua* Sunni or *qua* Shi'i. As such, it is as much a political identification as a metaphysical one, and one that pushes individuals toward greater extremes of what are already confessionally defined attitudes toward the Bahraini state.

The political connotations of sectarian identification are in evidence throughout the Bahrain survey, yet nowhere more so than in the case of national pride. Here one catches a rare empirical glimpse into the nature of the country's Sunni-Shi'i conflict; of the way that the mythical Shi'a-governed land of Ancient Bahrain competes in the popular imagination as in the political arena for supremacy and legitimacy over against the modern Kingdom of Bahrain conquered by "Al-Fatih" Ahmad bin Muhammad Al Khalifa. The data show that the same religious-cum-political identification that helps determine citizens' orientations toward the state also influences the amount of pride they feel for being Bahraini—that is, for being a "true" Bahraini. Although it drives Sunnis and Shi'is farther apart in their political opinions, now religiosity pushes them in the same direction toward greater national pride, albeit toward two separate prides representing two competing conceptions of nationhood.

Whereas some Gulf populations offer little variation in response to this standard survey item,[25] among Bahrainis considerable variation exists not only between the two confessional groups but also within them. The aim, then, is not to decide which of the communities is the more patriotic—a question made the subject of no little debate—but to observe the interplay between Bahrainis' sectarian and national identities. Respondents were asked directly, "How proud are you to be a Bahraini?" Almost four-fifths of Sunni respondents reported the highest level of national pride, compared to around 55 percent of Shi'a. Two things may help account for this not inconsiderable difference. The most obvious is a possible disproportionate inclination among Sunnis to feel (or convey) a stronger bond to the country and the larger political regime, in which, as we have seen already, they have a larger material stake than their Shi'a compatriots. The second likely cause is more subtle and turns on the acute politicization of the standard demonym "Bahraini," the necessary use of which in the survey questionnaire almost certainly served to alter the answers of some Shi'a who choose to identify primarily as Baharna. In fact, several Shi'i respondents specifically qualified their answer in this manner, saying, "I'm 'very proud'—not to be Bahraini, but to be *baḥrānī*!" Whatever the case, to interpret the between-group discrepancy as a deficit in national allegiance among Shi'a would be an oversimplification.

Far more interesting is the way that in-group identification interacts with feelings of national pride among all Bahrainis. Depicted in Figure 5.8 is the predicted likelihood that a respondent reports the highest level of national pride, disaggregated by sectarian affiliation and religiosity and while controlling for all other relevant variables. Among both Sunnis and Shi'a, being a religious individual is strongly associated with greater national pride. A religious Sunni is an estimated 21 percent more likely to be "very proud" of his nationality, in relative terms, than a non-religious Sunni. For Shi'a the boost is even more marked: from an estimated 40 percent among non-religious individuals to 62 percent among religious Shi'a, *ceteris paribus*, a relative increase of 54 percent. Both estimated effects are robust at the standard 0.05 level of statistical significance. Across Bahraini society, stronger in-group attachment is linked with stronger attachment to the country generally, nationalism having in this way a distinct confessional character tied to competing political understandings of what precisely Bahrain is and represents. Thus it is, for instance, that the country plays host annually to two separate National Days. An official holiday, on December 16, celebrates the coronation of Emir 'Isa bin Salman following Bahrain's formal independence from Britain in 1971. But, criticized as celebrating Al Khalifa rule more than the nation itself, the official date is shunned by many opposition-leaning citizens, who instead hold rival festivities on the actual day of independence, August 15.

Beyond shaping Bahrainis' political opinions and perceptions, personal religiosity also exerts a consistent and sizable influence over political behavior, boosting activism and participation among both Sunnis and Shi'is. Figure 5.9 summa-

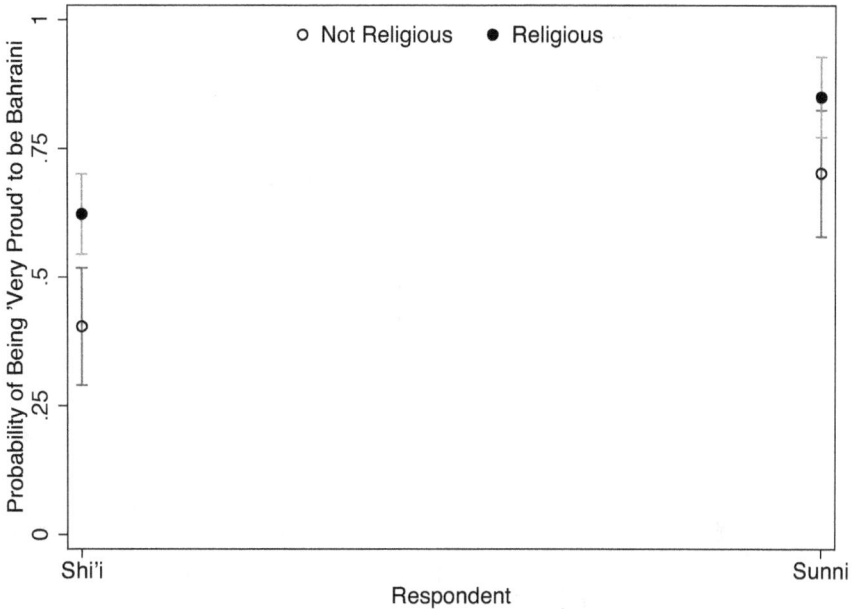

Figure 5.8. Religiosity and national pride.

rizes the results of three separate probit estimations investigating Bahrainis' participation in, respectively, street demonstrations, petitions and political meetings, and the 2006 parliamentary elections.[26] In each case, the data show an increased likelihood of participation among religious individuals corresponding in most instances to a substantively significant relative change. The predicted probability of having participated in a demonstration in the previous three years is 40 percent among non-religious Shiʻa, for example, compared to 54 percent among religious individuals. If similar in absolute terms, the change in probability among Sunnis from 11 percent to 21 percent corresponds to a substantial relative increase: a religious Sunni is almost twice as likely to have participated in a demonstration than a more secular individual. The findings are similar with respect to Bahrainis' involvement in political meetings and petitions. Here a religious Shiʻi is, again in relative terms, an estimated 89 percent more likely to have signed a petition or attended a political meeting, a religious Sunni 73 percent more likely. Finally, perhaps expectedly given the confessional bases of Bahrain's major political societies, religiosity also augments electoral participation, particularly among Shiʻa.

The question of participation in Bahrain's 2006 elections, one will recall, precipitated the fissure of al-Wifaq, with the offshoot al-Haqq Movement leading the charge for an electoral boycott. Not to be outdone, al-Wifaq promulgated a religious directive from Ayatollah ʻAli al-Sistani in which he ostensibly compelled pious

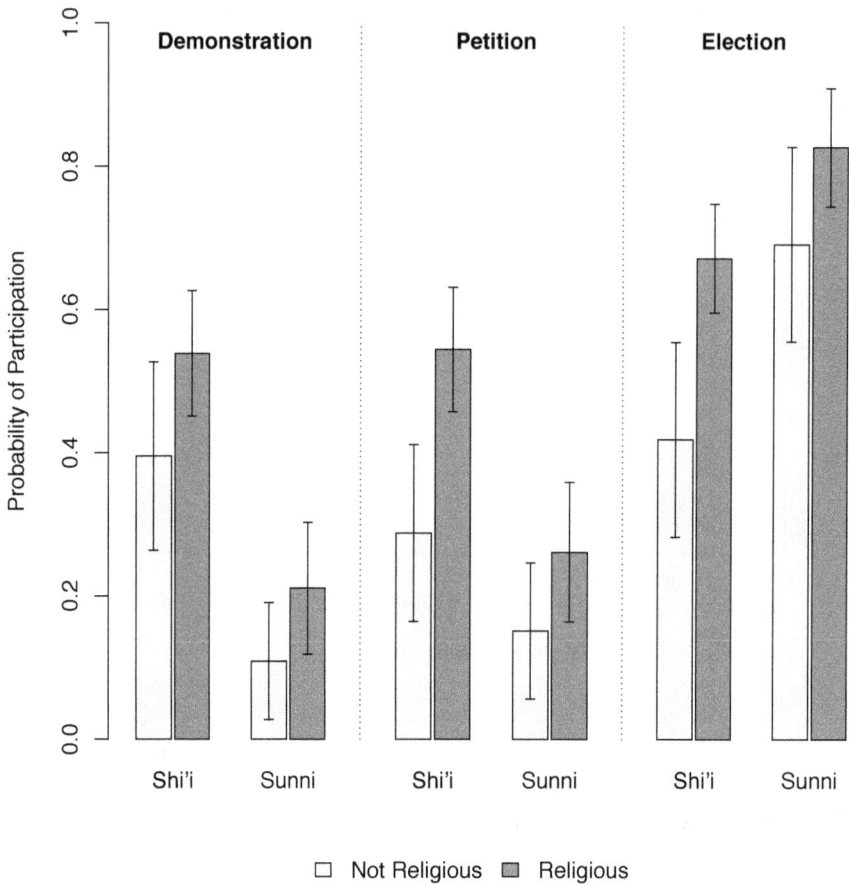

Figure 5.9. Religiosity and political behavior.

Shi'a to vote. As it would be the first election with official Shi'i participation, Sunnis were urged for their part not to be lulled into complacency by the prospect that many Shi'a would spurn the elections. The potential harm of not voting, according to Sunni religious leaders, outweighed that of government co-optation. The survey findings reflect this history. Sunnis have a higher probability of electoral participation in 2006 than Shi'a irrespective of religiousness. On the other hand, the relative impact of religiosity is much stronger among Shi'a, for whom being religious increases one's likelihood of participation from an estimated 42 percent to 67 percent, or by 60 percent. For secular Sunnis, the probability of voting is an estimated 69 percent, compared to 83 percent for religious individuals. The relative impact among Sunnis, then, is only around 20 percent.[27]

The Price of Allegiance

Yet, however instructive, these results cannot tell the complete story of the 2006 elections. To be sure, in lobbying their respective constituencies to take part in the vote, all but explicitly secular societies appealed to and capitalized on Bahrainis' confessionally informed political aspirations and apprehensions. So too, though, did these parties make other arguments based on more material concerns. A united Shi'a bloc in parliament, al-Wifaq vowed, would raise the standard of living for all Shi'a; whereas boycott would only ensure a greater socioeconomic divide as resource allocations were decided by a Sunni-dominated legislature. Sunni representation in parliament, the two Sunni Islamic societies told followers, would guarantee not only that "miscreants" could not "enact or pass laws incompatible with religious values," as al-Ma'awdah so instructively warned four years earlier, but also that an al-Wifaq majority could not channel state resources to its Shi'a supporters at the expense of ordinary Sunnis. Beyond the clear influence of sectarian membership and identification, therefore, to what extent did these economic-based arguments also convince Bahrainis to go to the polls?

The answer is that, if the appeals had any effect at all, it was upon Sunnis only. The estimated marginal effect of economy on the probability of electoral participation is virtually 0 among Shi'a respondents and around 10 percentage points among Sunnis. The probability that a Sunni of "very good" economy voted is an estimated 66 percent, compared to 77 percent for those of "good" economy, and 88 percent for those who rate their household economy as "poor."[28] By contrast, voter participation among Shi'a is estimated at around 60 percent regardless of economic circumstances. In deciding whether to take part in the 2006 elections, Sunnis were thus driven simultaneously by sectarian mistrust and their wallets; Shi'a, it would seem, only by the concerted appeal to their religious conscience made by al-Wifaq—and, eventually, by 'Ali al-Sistani's *fatwa*. The charge of al-Haqq leaders, that al-Wifaq and its clerical authorities "coerced" Shi'a to vote, here finds empirical evidence in the survey data.

Another way one may approach this question is by looking at respondents' reported political affiliations. Although too many individuals, especially Shi'a, declined to answer this question to allow a full regression analysis, still the data are instructive. Of those 127 Shi'is who did identify with a political society, a little over half, 55 percent, named al-Wifaq. A further 10 percent of respondents aligned with socialist-leaning Wa'ad, which attracts Sunnis together with Shi'a. Around 15 percent of Shi'is, finally, mentioned various minor groups, including local charities, human rights organizations, and liberal parties. The remainder, approximately 20 percent, identified al-Haqq. The average rating of household economy among al-Wifaq supporters and those who identify most closely with al-Haqq is not only statistically indistinguishable but indeed identical to three decimal places.

Bahrain's most ardent political opponents—those who, as it is said, refuse to "participate in the political process"—are not simply its poorest. Their stance of deliberate disengagement from the regime reflects a specific political orientation, not some underlying socioeconomic cause.

The overall finding with respect to Sunnis is no less compelling. In line with the *a priori* expectations stemming from our introductory examination of public benefits, here Sunnis are seen to exhibit considerable political sensitivity to economic conditions. Voting likelihood decreases as household economy improves, to the point that among those of the highest economic echelon no statistical difference in electoral participation separates Sunnis and Shi'is. Whereas, once again, the decision to boycott or participate among Shi'a is above all a matter of principle, Sunnis seem to aim to improve their own material circumstances by petitioning their benefactors directly. Among poorer Sunnis, then, there is no promise of remaining apolitical or politically supportive: if the state fails to uphold its end of the implicit bargain, why should Sunnis honor theirs?

This result is mirrored across a wide cross-section of political life. From general ratings of the government's political and economic performance to views of corruption to participation in important political activities, Sunnis reveal a vastly disproportionate sensitivity to economic conditions compared to Bahraini Shi'a. Table 5.4 reports the estimated relative change in eight different political indicators associated with a (hypothetical) change in household economy from "bad" to "good," and from "bad" to "very good," respectively, for Sunnis and Shi'is. For Shi'a, improved economic circumstances correspond to more positive political orientations in less than half of the cases—and in one of these, only for the more drastic change from a bad to very good economy. Moreover, this effect operates only in the realm of opinion, influencing neither Shi'a participation in demonstrations nor, as examined already, electoral participation.

Meanwhile, better economic conditions among Sunnis correspond to more positive political orientations almost universally. Across a wide swatch of political opinions and actions, improved household economy exerts a decided influence on Bahraini Sunnis, augmenting what are already (relative to Shi'a) positive orientations toward the state, and markedly decreasing the probability of non-sanctioned behaviors such as involvement in rallies and demonstrations. Better material conditions also boost Sunnis' national pride and, as among Shi'a, dampen support for radical versus more gradual political change. Notably, all but one of these effects obtains even for a single-category improvement from "bad" to "good" economy.

Consider the case of Bahrainis' participation in demonstrations. Overall, only 17 percent of Sunni respondents reported partaking in a demonstration or march in the previous three years, compared to more than half (52 percent) of Shi'a. Such a finding comes of course as little surprise, and fits well with qualitative impres-

Table 5.4. Bahrainis' political sensitivity to economic conditions

	Marginal Effect[a]			
	Shi'i		Sunni	
From a "bad" household economy to	Good	Very Good	Good	Very Good
Evaluation of				
Influence of policy on daily life	+18%*	+13%#	+24%**	+36%***
Government's economic management	—	—	+24%***	+41%***
Bahrain's overall political situation	—	+23%#	+14%*	+25%**
Prevalence of economic corruption	—	—	–13%#	–16%*
(1 least, 4 most)				
Agreement that				
"Reform should proceed little by little, rather than all at once."	15%**	21%**	+12%*	+17%**
(1 strongly disagree, 4 strongly agree)				
Participation in				
A demonstration	—	—	–59%#	–80%*
The 2006 parliamentary election	—	—	—	–23%*
National pride	—	—	+29%#	+47%**

[a] Post-regression linear predictions calculated using margins in Stata 12; evaluated at overall means of all other regressors. Predictions for "Participation in" category represent probit probabilities; those for the final category represent probabilities that a respondent is "very proud" to be a Bahraini. Significance levels associated with contrasts of predictive margins.
$p \leq 0.1$, * $p \leq 0.05$, ** $p \leq 0.01$, *** $p \leq 0.001$ (two-tailed)

sions about the relative political activism of Bahrain's sectarian communities at least since the 1990s Shi'a intifada. What this finding does not support, however, is the standard narrative, only further crystallized in the post-2011 period, of impoverished Shi'i opponents versus enriched Sunni loyalists. Indeed, the survey data show that precisely the opposite is true: it is poorer *Sunnis* who tend toward activism, not poorer Shi'a. As depicted in Figure 5.10, variation in economic circumstances plays utterly no role in influencing the likelihood of participation among Shi'a. Poorer Shi'a, it turns out, are no more likely to demonstrate than are any other Shi'a. The same cannot be said of Sunnis. The likelihood of participation for a Sunni of "very good" household economy is a mere 7 percent, all else being equal, but doubles to 15 percent among those of "good" economy; it jumps again to 35 percent among those of the lowest economic circumstances. Sunnis of the weakest economic group, then, are some five times more likely to have demonstrated

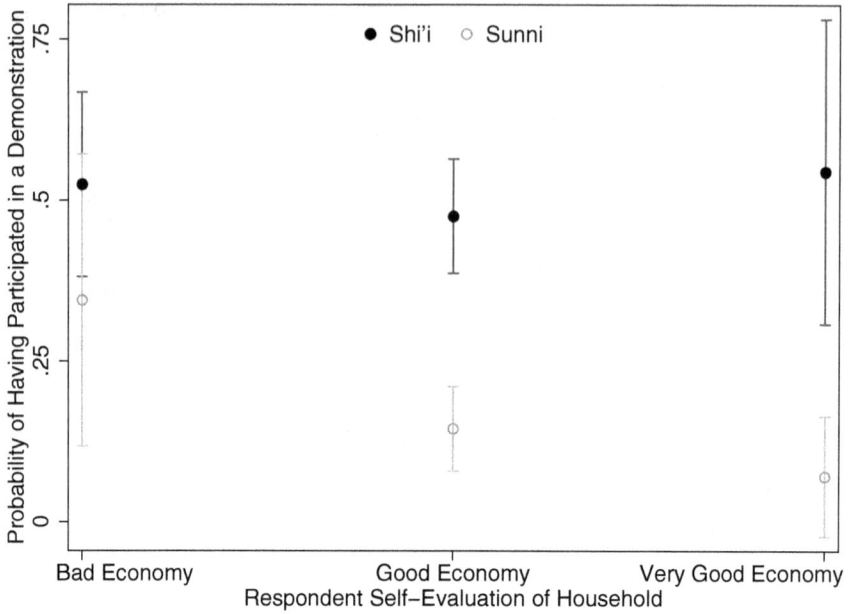

Figure 5.10. Probability of demonstration participation, by household economy.

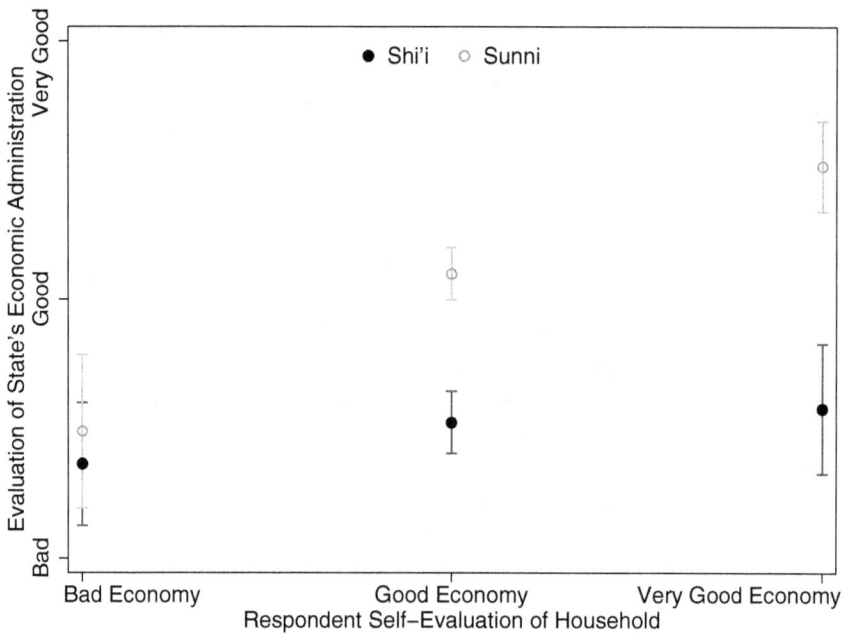

Figure 5.11. Household economy vs. rating of state's economic administration.

than the wealthiest group, an effect so strong that it erases the statistical difference in probability separating Sunnis and Shi'is of the lowest economic category.

Here again is clear evidence that the political views and behaviors of Bahrain's two confessional communities are shaped by fundamentally different processes. Among Shi'a, actions and attitudes are determined above all by one's larger ideological orientation toward the state rather than one's satisfaction with the material benefits afforded by it; whereas for Sunnis the process seems to follow more closely the economic-based mechanisms thought to link citizens to governments in the Arab Gulf context. A final plot illustrates perfectly this difference. Shown in Figure 5.11 is a respondent's view of the government's administration of the economy, compared to his own economic status. Among Sunnis the relationship is exactly as one would expect, with respondents rating the government more or less in accordance with their own economic experience: those doing well think the state is doing well; those worse off give the state a more negative evaluation. However, one can see that among Shi'a this relationship breaks down, or rather fails to materialize at all. Instead, Shi'a offer uniformly negative evaluations of the state's economic administration independent of their personal situation. In fact, the predicted rating of a Shi'i of the highest economic category is indistinguishable from that of a Sunni of the *lowest* category. For Shi'is, the question of the government's economic performance inescapably taps into larger perceptions regarding equality, fairness, and legitimacy—not simply in relation to the economy but to the state itself.

Containing the Sunni Awakening

Yet, out of place though it is within the typical storyline focusing on Bahrain's Shi'a-dominated opposition, the close link between politics and economics among the state's nominal Sunni support base is not something that finds evidence only in the mass survey data. In interviews, Sunni parliamentarians from both major societies described their legislative aims in expressly economic terms, often questioning what they perceived as the state's disproportionate political and financial concern for the very people who are against it. When asked about the origins of Shi'a frustration in Bahrain, the lawmakers retorted that, in fact, it is the Sunnis who have equal or greater cause for complaint. "[The Shi'a] villages used to be not cared for and were very backward and under-developed," noted Dr. 'Ali Ahmad, then and current leader of al-Manbar al-Islami. "Now a majority of the [government's] projects—in housing, for example—they are targeted toward the village areas." Another representative of the Muslim Brotherhood-affiliated society, Samy Qambar, made the same observation, saying,

> There is an area [in my district] of Riffa known as "Lebanon," which is one of the poorest regions of Bahrain. So it not simply that the Shi'a are poor and Sunna

rich. The challenge of poverty and socioeconomic inequality [in Bahrain] is not just a Shi'a issue or an issue based on religious differences. We [al-Manbar al-Islami] are working in the parliament to the raise the quality of life of these people as a whole—not just Sunnis or just Shi'a.[29]

The most emphatic response, however, came from Salafist parliamentarian 'Isa Abu al-Fath. "Look," he began,

the Shi'a need to understand that none of the GCC governments pay attention to their publics—it's not just them who are ignored. Even for us Sunnis—who represents us in the government? The Shi'a—at least they have [Muhammad 'Ali bin Mansur] al-Sitri, who is a special advisor to the King [for legislative affairs]; they also have several [Shi'i] ministers. Who do the Sunnis have? We are the ones suffering—more than them.[30]

This interpretation is not uncommon among Sunni and Shi'i alike. It follows an oft-heard Bahraini view that positions the royal family, and by extension the government, not as the unwavering benefactor of Sunnis but as the final arbiter—if not a neutral one—of the island's intrinsic societal dispute. The Al Khalifa are in this way seen as representing a "third sect," which in fact they are insofar as they and other tribal families alone follow the Maliki tradition of Islamic jurisprudence. Khuri, writing in 1980 of Bahrain's first-ever parliamentary elections nearly a decade before, noted:

Like all other ruling families in the Gulf and Arabia, the Al-Khalifa of Bahrain consider government a legitimate right that they earned historically by defending the island against external aggression—a "right" that must not be subjected to "the fluctuating, controversial moods of public opinion," as one Al-Khalifa sheikh put it. Members of the ruling family were not permitted to run for election because they were aloof from politics, above the National Assembly and the appeal to public opinion.[31]

The intervening three decades have done nothing to change this position. Despite sustained unrest, a 15-year constitutional reform movement, and an unprecedented popular uprising, the ruling family continues to deny that opposition grievances stem from actual government policy, insisting instead that the nation's essential conflict is between Sunnis and Shi'is themselves. Justice Minister Khalid bin 'Ali Al Khalifa in particular has given frequent voice to this description of the ruling family as unwitting intersectarian referee. During preelection turmoil in 2010, for instance, he told *The Economist* that the family positions itself as "'a buffer zone' between Sunni and Shia."[32] Beginning in early 2013 this ostensibly apolitical role was made even more explicit, with the justice minister chairing a series of farcical biweekly meetings billed as a "national dialogue." Government officials organized and "moderated" the talks between representatives of opposition and nationalist societies, while refusing to take part as an actual interlocutor. Predictably, the dialogue failed to make serious progress, in large part due to opposition insistence upon more serious high-level government involvement.

Hence, when 'Isa Abu al-Fath laments that "we Sunnis" have no one in the government to represent "us," he is not being disingenuous but simply excludes as a matter of course royals and political elites from traditionally allied families. These individuals, while denominationally Sunni, are identified above all as being of the ruling, tribal class, and, as such, they are seen to represent Sunnis no more than does Muhammad al-Sitri. Ordinary Sunnis, Abu al-Fath says in essence, are poorly repaid for the allegiance they show the Al Khalifa: disproportionately supportive of the government, they lose out on the majority of its benefaction to the very side that opposes it. This follows, Sunnis understand, not from malice but from the dictates of political expediency, which say that when resources are scarce, better to spend them where they are likely to matter most. If Sunnis can be expected to remain supportive of the status quo, whether due to natural disposition, out of sectarian affinity, or so as not to give ground to their political rivals, why then offer them resources that might be used to win additional friends from among today's enemies? When one is the Republican Party nominee for U.S. president, what purpose spending campaign dollars in Utah?

In this respect, arguably the most striking outcome of the February 2011 uprising was the sudden explosion, not of Shi'a anger, which was entirely predictable, but of this pent-up political frustration among Sunnis. Sidestepping established Islamist societies, a new generation of activists capitalized on the mass Sunni mobilization organized initially in support of the state to begin to advance an independent political program of their own. In addition to calls for harsher security measures against protesters, these new Sunni coalitions, incubated under the banner of the populist Gathering of National Unity, demanded a more efficacious role in decision making and a larger share of state resources. Among the most vocal, if ultimately short-lived, of these nascent coalitions was a TGONU offshoot called the Al-Fatih Youth Union, whose Friday rallies drew thousands of Sunni supporters for several months between late 2011 and early 2012. In December 2011, a sympathetic columnist at *Al-Watan* described the group as "a gathering of young people who do not want to be associated with any existing political trend," continuing ominously,

> They are fed up with the fact that those who have always supported the entity of Bahrain, Arabism, sovereignty and the royal family are being fooled because their loyalty is taken for granted; therefore they are treated as a reserve division. These are serious mistakes, and we will never know what they will lead to.[33]

As 2012 wore on, the nascent movement sputtered under coordinated pressure by the state. Vocal Sunni critics were targeted for questioning and, in at least one publicized case, were arrested.[34] In April, the business of prominent Sunni gadfly and independent MP Usama al-Tamimi was riddled with 30 bullets after he suggested in a session of parliament that the prime minister should be removed on account of corruption.[35] At the same time, the government moved to appease the less worrisome half of its Sunni detractors, namely those security-minded citizens advocating a redoubled crackdown on opposition activists. Arrests of high-profile protest leaders mounted, headlined by the detention in May 2012 of long-time pro-government target Nabeel Rajab, president of the Bahrain Centre for Human Rights and constant organizer of street protests. Previously off-limits leaders of al-Wifaq, including deputy leader Khalil al-Marzuq, Sh. 'Ali Salman, and Sh. 'Isa Qasim, were also made fair game. By the end of 2013, the former figures had been arrested and charged with various offences stemming from public statements critical of the regime. In January 2014, the Ulama Council of Shi'i scholars headed by Sh. 'Isa Qasim was banned for its alleged involvement in politics and failure to register with the Ministry of Justice. The ministry routinely threatened to extend the same treatment to Sh. 'Isa Qasim himself.

The paradox, then, is clear to see: Bahrain's sectarian strategy of governance, prompted by its practical inability to co-opt both Sunna and Shi'a simultaneously, has in the end failed to buy off either. Shi'a rightly perceive that their community receives a disproportionately small share of public and private benefits compared to Sunnis, and see in this evidence of unfairness, discrimination, and worse. Sunnis, on the other hand, though in actuality they reap an oversized portion of benefits, perceive instead a state constantly preoccupied with the demands of underserved Shi'a, a perennial crisis management that gives a skewed view of the government's priorities and distribution of resources. Even if they are happy to trade Al Khalifa leadership for that of al-Wifaq or its more radical contemporaries, Sunnis nonetheless harbor grievances of their own for which they are prepared to fight politically. While it is true that they can be brought back into line more easily, both because they stand to lose more from a change in the politico-economic status quo and due to the power of sectarian arguments about Shi'a and Iranian emboldening, still the resulting acquiescence is not an indication of positive political support, but of negative fear of change.

Bahrain's Chicken and Egg Problem

The Bahrain mass survey tells a simple, if in many ways uncomfortable, truth: the nation's Sunnis and Shi'is have, on the whole, fundamentally different political orientations and motivations that translate into substantively divergent views and actions vis-à-vis the state. Although it is perhaps reassuring to treat this outward sectarian difference as a mere epiphenomenon of some more mundane socio-economic disparity, the foregoing analysis has demonstrated definitively that such an understanding simply does not match reality. Sunnis tend to remain ideologically-supportive of the government *qua* protector of the status quo, even as they register their political grievances about economic conditions and other issues such as corruption and rampant naturalization that bear directly on the distribution of public housing, jobs, and other entitlements. Shiʿa remain opposed to the political status quo on principle, a position perhaps aggravated by but in the end independent of material circumstances. These confessionally-determined positions only harden further as in-group identification increases.

The survey findings therefore confirm each of the four main theoretical hypotheses outlined at the outset of this section. More than that, they are consistent with the results of the preceding examination of employment and other public benefits. Sunnis maintain a relatively constructive relationship with the state, yes, but not on account of altruism or sectarian affinity. The ruling family does not earn a free pass. For their near-unwavering support, and for their help in keeping the government's fiercest critics at bay, ordinary Sunnis expect—and have shown themselves willing to demand—something in return. That they enjoy a disproportionate share of the state's economic subsidies, therefore, is simply the government upholding its half of the unspoken agreement. Insofar as there exists in Bahrain a rentier bargain, this is it.

On the opposite end of the spectrum lie Bahraini Shiʿa, who as a group are party to no such agreement. Shiʿa citizens protest, vote in elections, and are generally oppositional not on the basis of economy, not because they seek redress for economic grievances, but on principle. Rather than material dissatisfaction, their political engagement and activism stems from dissatisfaction with the regime as a whole, wherein they find themselves limited as a group in political power and social standing on the basis of hereditary accident. Only one crucial matter remains unresolved, then, yet it is outside the reach of any survey data. It is the question of the origin of this negative political orientation among Shiʿa. Do Shiʿa form the core of political opposition in Bahrain because they are marginalized, or are they marginalized because they are viewed as intrinsically oppositional? In short, which came first: Shiʿa opposition or Shiʿa marginalization?

From a historical perspective, at least, the answer is clear. Bahrain's native Shiʿa inhabitants were subjugated politically and economically in the aftermath of the 1783 Al Khalifa conquest, a yolk lifted only forcibly via the British-backed administrative reforms of the 1920s. Although the indigenous population can hardly have welcomed the arrival of foreign tribal conquerors, still their domination came not fundamentally as punishment for opposition to the new regime, but from fiscal necessity (reinforced no doubt by notions of ethnic superiority). With the ruling Sunni class content to serve only as feudal landlords, and preoccupied in any case with internal and external tribal conflicts, Shiʿa cultivators and indentured pearl divers afforded the bulk of what little tax revenue the Al Khalifa could capture. It was only when the colonial administration altered its decades-long *laissez faire* policy toward Bahrain, seeking to replace the feudal estate system with more modern political and economic institutions, that 150 years of Shiʿa resentment exploded in the form of organized support for a British plan that would curb the state's absolute authority.

The ensuing conflict between Shiʿa supporters of the reforms, royal rejectors, and British-backed Al Khalifa moderates was protracted and bloody. It ended only after an emir's forced abdication, the banishment of obstructionists within the ruling family and allied tribes, and the death of many Shiʿa villagers at the hands of the latter.[36] Nonetheless, the previously unseen Shiʿa political mobilization had no roots in confessional identity per se, but in a century and a half of servitude and discrimination. The divide, if partly along sectarian lines, turned around a *political* question, which was whether one supported (or could accept) limits on tribal-cum-state power or, instead, viewed the British intervention as an intolerable encroachment upon traditional tribal prerogative.

Over the past four decades, however, the underlying conflict of Bahraini politics—or rather, the perception of that conflict—has shifted fundamentally. Primarily as a result of changing power dynamics and threat perceptions in the Gulf region and Levant, rather than any qualitative difference in the mode or substance of their political agenda, Bahraini Shiʿa are seen to be intrinsically oppositional and even disloyal, as if hostility to secular and Sunni-led government were genetically programmed. Ample evidence is seen in the Iranian Revolution and especially the post-2003 Shiʿa takeover of Iraq, to say nothing of the historical dispute over succession in the early Muslim community. Even localized movements, such as the Zaydi-turned-Twelver Huthi separatists—now effective rulers—of northern Yemen, are interpreted through the lens of Iranian-inspired Shiʿa agitation. Counterexamples, including ʿAlawi support for the secular Baʿathist regime in Syria, are assumed to reveal only the hypocrisy and sectarian character of the transnational Shiʿa agenda.

The upshot is that, for Bahrain and other Arab Gulf states nervous over expanding Iranian influence, an emboldened Shiʿa community is no longer a mere political problem to be dealt with using the standard incentives and tools available to a state, but a veritable problem of national security. If Shiʿa citizens are motivated only by a collective thirst for political power in order to rectify perceived historical injustice, then their activism cannot be dampened with promises of material reward. In that case, an economically well-off Shiʿi is not a more politically quiescent Shiʿi but simply a rich unhappy Shiʿi. Rather than waste resources trying to bribe a community intrinsically opposed to it, therefore, the ruling family has come to understand that it can use the Shiʿa community's capacity for group-based mobilization against it, by convincing ordinary Sunnis that it represents, more than a political threat, a threat to the very nation itself. Faced with the alternative of a clerical Shiʻi regime in Bahrain, most Sunnis will be willing to fight the government's battles for it rather than take the existential risk of joining forces with an opposition with which it may share many grievances, but that it does not trust to stick to a moderate agenda.

This "protection-racket politics," as Daniel Brumberg aptly describes many post-Arab Spring societies, is a common feature of Arab autocracies generally, many of which, as he says, display "an uncanny knack for manipulating a wide array of ethnic, religious, and sociocultural groups by playing upon their fears of political exclusion (or worse) under majority rule and offering them *Godfather*-style 'protection' in return for political support."[37] Both the empirical results examined in this chapter, as well as the story of the post-February 2011 period, suggest that Bahrain's capacity for sectarian manipulation is a political tool at least as powerful as its capacity to enrich.

6 Political Diversification in the Age of Regime Insecurity

In LATE APRIL 2011, six weeks into a brutal period of martial law that effectively ended the existential threat to the regime posed by the February 14th uprising, Prime Minister Khalifa bin Salman took the opportunity to thank the Bahraini people—government supporters, that is—"for their honourable mobilization against wicked plots," and "for standing united as a bulwark defending their country against subversive conspiracies." Speaking on state radio, he said, "The recent unrest has revealed the genuine mettle of citizens and revealed to the world the unity between people and the leadership in times of adversity."[1] For the premier, the affection was both genuine and personal, as his political head was atop the list of demonstrators' demands.

The month prior, Bahraini Sunnis had followed protesters into the streets— not to support the Shi'a- and secular-led movement but to help ensure it did not gain further traction. Mass pro-government rallies based at the Al-Fatih Mosque, namesake of the ruling family's famed conqueror, promised to preserve "al-baḥrayn al-khalīfiyya": Bahrain of the Al Khalifa. Before long this counterrevolution, which rivaled in numbers that of the opposition camp at the Pearl Roundabout, took on a more confrontational character, with armed loyalist mobs standing side by side with riot police in skirmishes with protesters. Sunni-Shi'i clashes at schools, on the campus of Bahrain's main public university, and in the streets threatened to culminate in open sectarian conflict.[2]

This breakdown in law and order, which mirrored a collapse in political negotiations between moderates in the government and opposition, helped precipitate the intervention on March 14 of the GCC's Peninsula Shield, a joint force consisting of several thousand ground troops from Saudi Arabia and smaller contingents from the UAE and Qatar. Within days, demonstrations were suppressed with renewed ferocity, a majority of opposition leaders were rounded up and prepared for military trial, and a three-month "state of national safety" was declared. The Pearl Roundabout was not simply cleared but razed to the ground, coins bearing its imaged removed from circulation. The uprising, so far as the state was concerned, was over. Yet, if the immediate cause was a redoubled police and military response, still there was no delusion about the decisive role played by ordinary Sunnis. A telling cartoon in the largest government-affiliated daily, *Akhbar*

Al-Khaleej, depicted a Bahraini ship once inundated by "the deviant sect" sailing again in calm waters, having been righted by the Gathering of National Unity.

A Body Divided

"Shiʻa opposition." "Shiite majority." "Shiʻa-led uprising." It is seemingly a matter of editorial regulation that political conflict in Bahrain be introduced with this sectarian qualifier. But such descriptions, if accurate, capture only half the story—and perhaps not the most interesting or important. The bloody and prolonged nature of the government-opposition confrontation, a political fire that has burned since King Hamad reneged on key reform promises with the 2002 constitution, continues to overshadow the other player—arguably the *key* player—in Bahraini politics, which is the state's nominal Sunni support base. While, especially in the post-2011 period, Sunnis have demonstrated an intermittent willingness to challenge the state on specific policies, their negative role as counter-opposition has far exceeded their positive pursuit of an independent platform. By and large, Bahraini Sunnis have been content to fight the state's battles for it, both in parliament and in the streets.

That decades have passed without the materialization of a successful cross-sectarian political movement in Bahrain is, of course, no accident. The state has a direct interest—and direct hand—in preventing such an emergence, as cross-societal coordination alone represents a viable political threat to the regime. Shiʻa can complain and protest; can even mobilize hundreds of thousands of like-minded opponents. But so long as they act alone, they lack the military preponderance necessary to physically overthrow the state, and their efforts will be in vain. Thus, it was no coincidence that secular leaders, and especially Sunnis who dared to "break ranks" to join demonstrators at the Pearl Roundabout, were singled out for swift retribution. This response was simply an extension of longstanding policy.

The country's most popular secular opposition group, Waʻad, had for years been the target of electoral manipulation. Despite a wide following and two charismatic leaders, concerted government-led mobilization of voters (including the alleged use of police and military personnel) in districts in which Waʻad fielded candidates ensured that the group did not gain a single seat in parliament in either election it contested. Even more dramatically, in 1975 the entire National Assembly was dissolved but two years into its first-ever term when an unlikely leftist-Islamist alliance was poised to overturn a draconian state security law. Khuri relates that authorities first attempted to win over the latter bloc by branding as "blasphemous" its socialist-oriented allies. When this religious appeal failed, the emir simply scuttled the parliamentary experiment altogether.[3]

Luckily for the present regime, the resurrection of parliament in 2002 came amid a changing domestic and regional climate in which such arguments would, especially after the U.S.-enabled Shiʻa takeover of Iraq and ensuing marginaliza-tion of Sunnis there, regain their force with redoubled potency. For most Gulf Sun-nis today, a stingy or authoritarian tribal ruler is a source of relative frustration and perhaps discontent; the specter of a turbaned clerical leadership not just in-tolerable but genuinely frightening. "If the Shias took control of the country," one plain-speaking Bahraini told the *New York Times* amid preelection turmoil in August 2010, "they would pop out one eye of every Sunni."[4] Its conspicuous presence in the traditional and social media gives the impression that the acute sectarian distrust that has characterized the post-2011 period is itself mainly a product of the uprising. But well before, the government's most important political message—you're either with us or with the ayatollahs in Iran—had found an attentive audience.

Indeed, its deleterious effects are palpable even in the mass survey data—not in the polarized views of citizens merely but in the very interaction between interviewer and respondent. The reader will recall that included in the empirical models of chapter 5 was an indicator capturing the confessional composition of the interview, that is, whether the respondent and interviewer shared the same sectarian affiliation. This control was meant as a statistical correction for the social desirability bias inherent in mixed-sect interviews. But beyond its meth-odological purpose, what this indicator tells about citizens' responses to sensitive political questions is revelatory in its own right, as it provides empirical-based in-sight into the nature of interpersonal trust—or, as is more often the case, distrust—in Bahraini society.

In short, individuals from both communities systematically misrepresented their political views when interviewed by someone of the opposite sect. More spe-cifically, Shiʻa interviewed by Sunnis tended to report more positive political opin-ions, while Sunnis interviewed by Shiʻa tended to give more negative, more Shiʻa-like evaluations. It is not simply, then, that individuals gave politically safer (i.e., more pro-government) answers in mixed-sect interviews owing to some increased discomfort or suspicion in regard to the survey as a whole.[5] Rather, respondents misrepresented their views to make them appear closer to the views they attributed, strictly on the basis of presumed sectarian affiliation, to their interviewer. The an-swers are in this sense indeed "safer," yet not safer vis-à-vis the state but vis-à-vis one's interviewer, to whom a respondent ascribes confessionally defined political views that he imagines differ exactly on this account from his own.

Consider, for instance, the case of Bahrainis' opinions of the 2006 parliamen-tary elections, marred by allegations of fraud and manipulation issued in the Ban-dar Report but a month prior to the vote. As illustrated in Figure 6.1, those re-spondents interviewed by members of their own confessional community report evaluations in line with those expressed at the time by the Shiʻa-led opposition and the government's Sunni-dominated support base. All else equal, the predicted

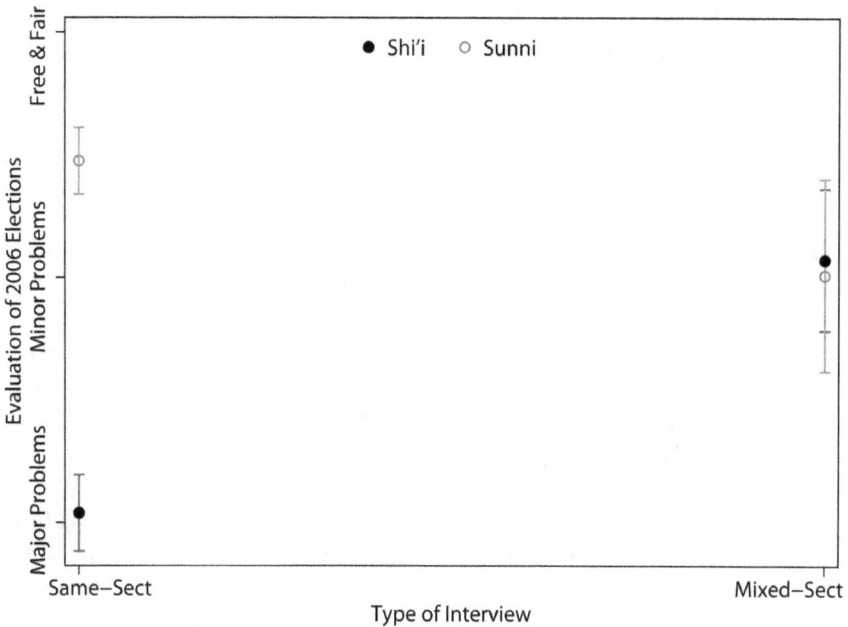

Figure 6.1. Evaluations of the 2006 parliamentary election, by interview type.

response among Shi'a is that the elections were affected by "major problems," compared to an evaluation somewhere between "minor problems" and "free and fair" for the average Sunni. Among mixed-sect interviewees, however, which include 27 percent of all Shi'i respondents and 17 percent of Sunni respondents, these expected views are tempered on both sides to the point that there is no longer any between-group difference in opinion. If one would look only at inter-sect interviews, indeed, there would appear to be no political conflict at all. This moderating effect, especially strong among Shi'a interviewed by Sunni fieldworkers, is systematic throughout the survey.[6]

It is a question for psychology precisely when and why an individual will tend to misrepresent his views to avoid confrontation with another who is believed to disagree. The notable upshot for our purposes is that the survey data offer evidence not only of a broad Sunni-Shi'i divergence in political views and behavior, but so too a wide disconnect in everyday societal interaction. Mixed-sect interviews affect respondents not by making them less comfortable about participation in the survey per se, but by making them more mindful of their opinion relative to the opinion they ascribe to their interlocutor. Thus, beyond its critical role in shaping political orientations and behavior, confessional division in Bahrain also serves to perpetuate discord through its impact on basic social interaction between citizens. Rather than reveal their true preferences to their political rivals, Sunnis and especially Shiʻa tend instead to misrepresent themselves as being less extreme and more conciliatory in their positions, giving the false impression that the two sides are perhaps not so far distanced politically and socially as they in fact remain.

Of course, the state's active cultivation of sectarian distrust tells only half the story of Bahrain's more general political polarization. The other cause, equally apparent in the survey findings, is that the two communities simply face two different sets of incentives. For Sunnis, political support can be expected to earn some earthly reward, and so is a relatively low price to pay independent of more intangible concerns over potential Shiʻa empowerment, Iranian interference, and so on. And, to the extent that this reward is not realized, Bahraini Sunnis are willing, albeit within a relatively constrained set of political bounds, to alter their behavior accordingly. For Shiʻa citizens, by contrast, whom the rentier welfare state serves disproportionately poorly, the expected marginal benefit of a more moderate political orientation is far less enticing—indeed, in many cases perhaps offers little more than could be expected from a life of political activism. Under such circumstances, the potential payoff of the latter, of seeking fundamental reform of the prevailing system, may so outweigh the alternative path that opposition accords not only with political principle, but so too with a citizen's evaluation of his or her own material interests.

The Trouble with Bahrain

This book has sought to resolve several longstanding empirical and theoretical puzzles surrounding political life in the resource-dependent societies of the Arab Gulf, problems further illuminated by, but fundamentally independent of, Bahrain's popular uprising of February 2011 and its far-reaching regional repercussions. Far from the notion of popular resignation from politics, in fact the Arab Gulf is not only home to citizens who take an active interest in politics, but also three of the six GCC states, and not simply the poorest—Bahrain, Saudi Arabia, and Kuwait—feature no less than organized political oppositions, each decades old. And yet, despite this surprising cache of public enthusiasm, somehow these Gulf regimes as a distinct class of state still have managed to enjoy greater political stability than other Arab countries. How can this be?

The conceptual issue attending these empirical observations presents no less of a problem. This is that, for a theoretical framework that has dominated the interpretation of Arab Gulf politics since its initial articulation some thirty years ago, the individual-level behavioral assumptions upon which rentier theory rests have remained all but unexplored. Before one can answer the question of why some Arab Gulf regimes appear unable to buy political support with economic benefits, it is necessary to know, in the first place, whether it is indeed material well-being that tends to determine the political views and behavior of ordinary Gulf Arabs, and, moreover, whether governments are uniformly willing and able to offer such a bargain. The rentier state hypothesis having sprung from economics, that rational, self-maximizing citizens and rulers would happily barter political privileges for tangible goods was presumed as a matter of course. And, with no individual-level data to suggest otherwise, eventually this thesis achieved the status of a truism, assisted by a convenient parallel to the Western experience that was soon to be pointed out: "the reverse principle of no representation without taxation." But such maxims are poor substitutes for empirical evidence.

The foregoing study, informed by previously unavailable sources of insight, has aimed to address these problems by answering three basic questions. First, what is it about Bahrain *qua* rentier society that renders its rulers particularly incapable of buying popular political quiet? Second, to what extent does the variable that explains the case of Bahrain in fact describe a larger class of rentier societies that share this causal feature? And, third, should it exist, in what ways must this latter category demand a revision of the theoretical framework that underlies the prevailing interpretation of politics in the Arab Gulf? Here we review our answers to these questions, note their limitations, and finally suggest an updated Arab Gulf research agenda that retreats from the economics-based model of the self-maximizing rentier state and citizen in favor of one rooted more firmly in the observed politics of the region.

Bahrain's inability to buy popular political disinterest and assent owes directly to society's broad division into ascriptive group constituencies. Beyond offering a viable basis for mass coordination in a type of regime whose very productive organization is assumed to preclude one, group conflict also serves to disrupt the mechanisms of political buy-off available to Bahrain *qua* rentier state. In the first instance, it renders the political orientations of ordinary Bahrainis dependent not primarily upon economics but upon more intangible factors such as confessional identity and support for democratic governance (or majoritarianism) as an inherent political good. At the same time, it induces the government to forego the liberal and indiscriminant deployment of state benefits to mollify would-be opponents, for fears over national security, of Iranian-inspired Shi'a emboldening, and, more generally, as (real or preemptive) punishment for those perceived to lack national loyalty. The latter effect, which operates on the political supply side, means that the state is unable to utilize effectively even those political pressure-relieving levers said to be distinctively available to it as an allocative economy. The former consequence, operating on the demand side, ensures that such state-side efforts at appeasement, even if they could be employed, would be employed in vain.

Hence, in Bahrain and in other societies divided along ascriptive lines, citizens' political orientations will depend fundamentally on their perceptions of the balance of power enshrined in the extant regime. Concerns for the empowerment of one's rivals at the expense of one's own group inevitably compete with more mundane matters of economic welfare in determining the extent of an individual's support for, and actions in favor of or against, the government as conservator of the political status quo. Political attitudes and actions are influenced not simply by the question "What has the government done for me?" but by the more elementary question "What—or who—does the government represent? By whom exactly am I being governed? And is it my interest they have in mind?" Bahrain's Shi'a out-group is oppositional on principle, owing to its structural exclusion from the instruments of power, not from dissatisfaction with its collective share of the nation's oil revenues. Thus, for example, anti-government demonstrators were only enraged further when King Hamad attempted to preempt their February 2011 protests with the announcement, three days before demonstrations were set to begin, of a 1,000 dinar (around $2,650) hand-out for each Bahraini family. As expressed by one now-imprisoned opposition leader, "This is about dignity and freedom—it's not about filling our stomachs. . . . The core of this is political, not financial."[7]

The issue is also, for many Shi'a, one for which inspiration may be found readily in religion itself, the historical arc of Shi'ism being precisely one of struggle and self-sacrifice in the face of a more powerful but corrupt political-cum-religious oppressor. In Bahrain, religious rites and celebrations are replete with allegory and even explicit comparison connecting the seventh-century conflict over the leadership of the Muslim community to the Shi'a community's present-day battles against perceived discrimination and authoritarianism. Indeed, to an outside observer of the annual celebration of Ashura, it is difficult to perceive whether the myriad processions, passions plays, and sermons tell of the battle with the Umayyad caliph in 680 CE, or the current struggle against the Al Khalifa dynasty. In sum, many Shi'a believe or can be motivated by the notion that they have a collective right to political authority based on notions of injustice and betrayal rooted in the very foundations of their religion; thus, the state is unable to pacify their demands with mere promises of jobs and relief from taxation.

At the same time, anti-state mobilization on the part of Bahrain's Shi'a-dominated opposition elicits a popular response in kind from ordinary Sunni citizens anxious to avoid any qualitative shift in the nation's political balance of power, much less a wholesale change in regime instigated by presumed Iranian agents. As a result of this Sunni countermobilization, in the divided context of Bahrain it is not only members of the political out-group that defy the basic assumption of apoliticality, but the entire community, pushed on the one side by Shi'a and allied secular reformists, pulled on the other by a countervailing force of Sunnis motivated by the fear, exaggerated or not, that a Shi'a-empowered Bahrain inevitably will begin down the course of post-2003 Iraq.

The rentier state of the Arab Gulf must therefore sink or swim on its capacity for economic appeasement; however, in divided societies this ability is hampered not only on the demand side by those citizens unwilling to take the bargain, but also on the supply side by a state reluctant to enrich or empower members of a community it views as an open or latent political opposition with ties to hostile regional challengers, individuals readily identifiable moreover on the basis of geography, family names, dialect, and other outward group markers. The question such a state faces, accordingly, is whether its power of economic benefaction is best used to reward friends or to attempt to convert known and potential enemies.

In Bahrain, at least, the answer is clear: it is political allegiance that secures material benefits, not the reverse. In case there were any doubts about the direction of causality here, one need only recall the thousands of Shiʻa dismissed from public-sector positions in early 2011 for alleged involvement in opposition activities. More recently, the state used the threat of similar economic repercussions to induce voter turnout for the November 2014 elections in the face of an opposition boycott. Just two days before the first round of voting, the head of Bahrain's High Elections Committee revealed publically that the cabinet "is studying procedures and administrative measures against those who miss out intentionally on the elections," warning that "options included preventing those who don't take part in the election from getting a job in government."[8]

The demand for political allegiance as a prerequisite for employment is even more stringent when the work in question carries national security implications. And, in a part of the world that spends more of its wealth on internal and external security than any other,[9] the scope of the resulting group-based exclusion from this most far-reaching of rentier benefits is far from trivial. Not only are Bahraini Shiʻa excluded altogether from police and military service, but fear of Iranian-inspired emboldening serves to limit their employment also in those institutions close to the exercise of state power, including the Ministries of Defense, Interior, Foreign Affairs, Justice, and others. And where they do find public employment, Shiʻa are suffered to fill lower-ranking positions than Sunnis of equal objective qualifications. Paradoxically, then, though with only economic patronage at its political disposal, still Bahrain—this failed rentier state—opts to forgo or curtail what is assumed its most powerful weapon, for fear that the cure should be worse than the disease.

Empirical Limitations

Each of these lines of argument found compelling evidence in the analysis of mass survey data from Bahrain in chapter 5. First, it was shown that Sunni citizens enjoy a decided advantage over their Shiʻi co-nationals both in securing work in the government sector as well as in the professional level of that work. More particularly, among citizens with no university education—representing approximately 45 percent of all Bahrainis as per the survey sample—a Sunni is far more likely to have a job in the public sector, conditional on employment, than a Shiʻi of identical employment-relevant attributes. Whereas a Shiʻi with a primary or lower terminal education is estimated to have only a 23 percent conditional probability of being employed in the state sector, holding constant the effects of age and gender, a working Sunni of the same education level is effectively guaranteed to have a government position. A similar gap exists for citizens with a terminal secondary education: a Shiʻi possessing a secondary school diploma is estimated to have just a 36 percent chance of working in the public sector, compared to a predicted 62 percent among Sunni high school graduates. The discrepancy in professional status is of similar substantive magnitude.

In addition, the data revealed, whereas 17 percent of working Sunni males who reported professional data indicated that they worked for the police or armed forces; and whereas 13 percent of all Sunni households reported at least one member employed in these services, not a single individual from among 117 employed Shiʻi males who offered occupational data reported working for the police or military. The patterns of government-sector employment in Bahrain thus tell a story fundamentally different from the one articulated by rentier theorists.

The survey analysis also confirmed the primacy of confessional membership over against economic status in determining Bahrainis' political orientations and behaviors at the individual level. Across numerous dependent variables capturing different aspects of political opinion and involvement, the most substantively and statistically significant predictor was a respondent's sectarian affiliation, and by no small margin. Moreover, the analysis found, Bahrainis are further entrenched in their respective confessionally defined opinions—Shiʻa tending toward more negative orientations, Sunnis toward more positive—by the additional augmenting influence of religiosity, used to proxy for the strength of one's confessional group identification. The influence of household economy, on the other hand, proved a reliable predictor of political orientations only among Sunnis, a result consistent with the notion that the citizen-state clientelism presumed to operate in Bahrain operates disproportionately among members of this community.

If the Bahrain survey thus provided strong evidence in support of the conceptual framework introduced in chapter 1 and given further substance in chapters 2 and 3, still this quantitative investigation was not without practical and methodological shortcomings. Before proceeding to consider the second question here about the larger applicability of these empirical findings, then, we might first pause to assess their limitations. Concerns regarding the actual survey procedure require perhaps little additional attention. Certainly it would have been preferable to have acquired a sample of more than 500 households, to have completed surveying of all these households, and to have achieved a more proportional representation of males and females among respondents. Yet, as noted previously, on account of

Bahrain's miniscule citizen population, the survey's 500-household sample in fact represents a sample-to-population ratio that surpasses that of any other Arab Barometer survey undertaken to date. Furthermore, as examined at length in chapter 4, the final geographical distribution of administrative units included in the sample conforms almost exactly to their relative national-level proportions according to two difference sources, one of which is the state's own 2010 national census.

As regards the latter points, indeed, interviewing reached only 435 of 500 sampled households, largely due to political and social tensions before and during the survey period. Still, because these remaining un-surveyed households were distributed randomly, there is no reason to believe that their omission compromised the representativeness of the final 435-observation sample. About the underrepresentation of female respondents, finally, in particular among rural Shi'a households, little can be said but that the practicalities of surveying Bahrain's isolated and conservative villages make such a result almost ensured. One might have oversampled female respondents, but such a procedure would have militated against the competing goal of reaching a maximum share of the 500 sampled households, most of which were found in hostile environments. In the end, at a time when any Bahrain mass survey at all seemed frequently in doubt, this and other concessions were made necessary.

Beyond sampling issues, one may identify shortcomings in the survey instrument itself, namely in the validity of the indicators used to measure the main theoretical concepts underlying the quantitative analysis: the independent variables measuring economic satisfaction and religiosity. These issues, noted where relevant in the foregoing chapters, may be summarized here. The first case is not cause for much worry but bears repeating nonetheless. This is that the indicator used to measure economic satisfaction among Bahrainis in fact was more closely an indicator of Bahrainis' household economy per se, respondents being asked to rate their household's financial situation from "very good" to "very poor." Moreover, the distribution of responses offered less variation than one would have liked, with about two-thirds of both Sunnis and Shi'a reporting a "good" financial situation and only a combined 2 percent describing their finances as "very bad." The latter category was accordingly collapsed with "bad" to create a three-category indicator. Still, since several other independent variables of interest—sectarian affiliation, religiosity, and the indictor signifying an inter-sect interview—were dichotomous measures with even less variation by their very construction, this concern should not be overstated.

One might also wonder about the measure of Bahrainis' religiosity. In the absence of a direct question about respondents' sectarian attitudes, which was deemed too sensitive to ask, the religiosity indicator served as a necessary proxy for a related but different concept: the strength of an individual's confessional identification as a Sunni or as a Shi'i. Questions about the theoretical validity of the religiosity measure were compounded by the choice of its actual coding, which presented two imperfect alternatives. A first was based on a straightforward query, but one that invited dishonesty and perhaps a Sunni-Shiʿi discrepancy in interpretation; a second involved yet another layer of proxy but thereby avoided manipulation and, it was thought, between-group inconsistency. The latter course was thus adopted.

A final topic we may treat briefly is the actual model specification used throughout the survey analysis. Compared to the alternative of segregated Sunni and Shiʿi models, the interactive specification offered most of the benefits of the former on top of additional advantages that only it could provide. Key among the latter are: utilization of the full sample of observations rather than two precariously small subsamples; the ability to estimate the critical effect of sectarian group membership itself, which is not possible using separate Sunni- and Shiʿi-only models; and, not least, greater correspondence to our conceptual argument, which does not ask, "Among Bahraini Shiʿis (or Sunnis), what are the factors that affect political opinion and behavior?" but rather, "Among ordinary Bahraini citizens, what is the independent effect on political opinion and behavior of, among other things, being a Sunni rather than a Shiʿi?" Finally, although these were not of major analytical concern, the interactive model also preserved group-specific estimates of the effects of the demographic control variables.

Revising the Rentier State Framework

From here it is obvious what revisions are necessary to the extant rentier state framework. A first regards the naïve assumption that economic benefits are distributed by resource-rich regimes in a manner that is politically agnostic. The cliché image of the government critic-turned-government minister found little empirical or even anecdotal support in the foregoing. Instead, the cause and effect would seem in Bahrain at least to be reversed: material benefits are not employed primarily in purchasing new political supporters, but in rewarding existing ones. Increased political deference on the individual level is thus a principal *cause* of receiving greater economic rewards from the state—not, as rentier theorists would have it, the effect.

Such a relationship is reinforced even further by the nature of the public sector in the Arab Gulf, in particular its ever-mounting securitization. In the decade between 2000 and 2009, five of the six GCC states counted among the top eleven military spenders as a proportion of GDP, and four fell within the top six.[10] Accordingly, the decision to extend a citizen employment in the state sector is one that often—and increasingly—intersects with national security concerns. To be sure, it is for precisely this reason that most of the region's militaries and security services are staffed largely with non-nationals assumed to have loyalty to no one but the state, with only officers taken from among the ruling and allied tribes.

Service in ministries involved in the exercise of state power demands similar precaution. For those innocuous civilian positions that do remain, preference goes to helping out political friends rather than buying off political enemies. And, in societies in which ascriptive identifiers are assumed, rightly or wrongly, to communicate not only ethnic data, but also information about an individual's probable political leanings, such targeted decisions to reward or punish are easily made. Beblawi insists that "[e]very citizen" of a rent-based regime "has a legitimate aspiration to be a government employee; in most cases this aspiration is fulfilled." While he fails to elaborate the opposite case, we may say that group-based societal division and political mobilization, combined with a near-hysterical fear of foreign-inspired irredentism, is the cause of one such exception.

Political life in the Arab Gulf states can no longer be summarized neatly and axiomatically as a pragmatic bargain of economic happiness for political quietude. For, even if it were true that material satisfaction engenders more pro-regime orientations on the margin, in the first place such an effect does not lead necessarily to *apoliticality* but, in contested societies, to an *increase* in political action in defense of a regime and of one's more favorable position therein as member of the benefited class. In the second place, economic incentives operate as but one of several competing determinants of political views and behavior. To restate an earlier formulation: that one's interest in politics does not stem from the wish to oversee the usage of one's taxed income does not mean that one is disinterested in politics. So too, there are many grounds upon which a Bahraini or other Gulf Arab may oppose or support the political status quo, and the relative fullness of his wallet is only one of them.

As for the larger question of how the Gulf region as a distinct category of states continues to enjoy relative stability compared to the rest of the Arab world, and this despite an ever-heightening political consciousness due to the effects of Sunni-Shiʻi division and the aggravating impact of regional geopolitical rivalry, it is plain now that the answer is not simply that the region's would-be political activists are paid by their governments to shut up. Instead, rather than seek to transform regime critics into regime clients, the ascriptively divided states of the Arab Gulf tend to cultivate in their place a more dependable ally: a captive ethno-religious constituency that already shares (or can be persuaded that it shares) an interest in preventing any significant change to the political status quo. Indeed, if the Sunni citizens of Bahrain will tend to support the prevailing political system irrespective of whether it benefits them personally, simply because they prefer it to what they imagine as the alternative, then why bother trying to win over Shiʻa citizens who, from the regime's perspective, will never surrender their true loyalties in any event? More dramatically, if hundreds of thousands of Bahraini Sunnis are willing to mobilize largely of their own accord in order to avert a perceived takeover by religiously inspired Shiʻa revolutionaries, why use one's limited resources to court the potential political support of the latter when it can be better spent in rewarding and thereby reinforcing the already demonstrated support of the former?

The Gulf state has, in addition, a final trick at its disposal. Faced with a rebellious population, it can simply procure a new one, by extending citizenship to those presumed to be more manageable and revoking that of opponents. The attraction is twofold: a diluted native citizenry and a newfound political constituency that owes its entire economic and political livelihood to the state. In the two decades following independence in 1961, for instance, Kuwait granted citizenship to more than 200,000 Sunni tribesmen,[11] first to help marginalize urban merchants and Nasserites, and later to dampen the electoral influence of Shi'a following the Iranian Revolution. Between 2004 and 2005, Qatar stripped the citizenship of between six and ten thousand members of the al-Ghufran branch of the al-Murra tribe, the largest in Qatar, for their alleged support for a Saudi-backed failed coup attempt a decade earlier. It was restored in less than a year for most of those affected, but not before many were forced into temporary exile and a powerful message had been sent.[12] In December 2011, the United Arab Emirates stripped the citizenship of seven individuals suspected of ties with the banned Muslim Brotherhood who signed a petition calling for the UAE's elected advisory body, the Federal National Council, to be given more powers.[13] A year later, Bahrain followed the Emirates' lead to strip Bahraini nationality from 31 Shi'a activists, including a former parliamentarian.[14] A royal decree issued in August 2013, ostensibly at the urging of pro-government legislators, expanded the grounds for the revocation of citizenship to all those who "help or serve a foreign country" or "endanger state interests."[15] And, in the first ten months of 2014 alone, Kuwait stripped at least 33 individuals of nationality following a cabinet-directed review of persons who "undermine or threaten the country's stability."[16]

As seen already, however, Bahrain's demographic machinations extend far further. Since at least the late 1990s, the ruling family has enacted a deliberate strategy of sectarian engineering, naturalizing Arab and non-Arab Sunnis for work mainly in the police and military sectors. Beyond building a largely non-indigenous—opponents would say "mercenary"—army and internal security service, the state's effort also has had profound political and social impacts. Military and police personnel have allegedly been used for vote-rigging at so-called general polling stations not tied to specific constituencies, while immigrants' ability to circumvent decades-long waiting lists for public housing and other benefits is a perennial source of aggravation for "original" Bahrainis of all backgrounds. More generally, both Sunnis and Shi'is decry what they see as a dilution and cheapening of Bahraini nationality, seemingly up for grabs for anyone capable of handling a weapon.

Yet, for the Shi'a-led opposition, the state's continued leveraging of citizenship as a political tool represents something still worse: the prospect that, as goes the community's demographic advantage, so too goes the force of its majoritarian claims to a greater share of political authority. If the story of Bahrain is no longer that of a Sunni autocracy repressing a disenfranchised Shi'a citizenry, but merely a Sunni-majority population content to accept the limited reforms volunteered by the state, then the narrative at least, if perhaps not the underlying political dynamic, is fundamentally changed. Even more important, insofar as the state is confident in having secured a Sunni majority, it can, if necessary, proceed down the path of actual institutional reform—including by meeting long-standing opposition demands such as fairer electoral districts, free parliamentary elections, and a more empowered legislature. The government could thereby diffuse political pressure while simultaneously disarming the opposition.

For, if Bahrain's Shi'a population is so marginalized numerically that it is impossible even in free and fair elections for opposition societies to capture the votes necessary for a legislative majority, and thus to shape parliament's composition and agenda, then the state can negate one of its detractors' most powerful arguments, namely that the majority viewpoint is not properly reflected in the country's professedly "democratic" representative institutions. In that case, continued extra-parliamentary activism is more likely to be viewed by once-critical observers not as legitimate demands for change but as unjustified and undemocratic protest from parties unwilling to accept having lost a fair electoral fight. Based on conversations with al-Wifaq leaders,[17] this scenario is more than of mere hypothetical concern.

Yet, opposition anxieties notwithstanding, one doubts whether such a strategy, attractive though it may be in theory, could ever be entertained seriously by Bahrain's rulers. In the first place, the state would require supreme confidence not only in its ability to engineer the demographic composition of electoral districts, but also in its understanding of electoral behavior, including both voter preferences (for example, Sunni and Shi'i support for Islamist vs. secular societies and candidates) and respective voter turnout among members of the two communities. Even apart from these procedural questions, moreover, the Al Khalifa would require in addition a heretofore undemonstrated *political* confidence in their own nominal Sunni support base: trust that Sunni politicians and societies, once elected and newly-endowed with substantive power, will remain satisfied with their traditional role as legislative obstructionists, and will not seek an independent, and potentially more confrontational, role in decision-making.

Indeed, if the unilateral electoral district changes announced in September 2014 are any indication, official concern over potential electoral and legislative coordination among Sunnis is today a motivating force even in the absence of an empowered parliament. Whereas the redrawn boundaries neither helped nor harmed al-Wifaq's electoral prospects[18]—a fact that aided in spurring an opposition boycott of the 2014 general elections on the grounds that the government failed to rectify sectarian-based gerrymandering—the new constituencies severely hindered the chances of Sunni Islamist and populist candidates in favor of tribal independents. Districts in the Sunni-dominated South were substantially expanded to include new neighborhoods belonging formerly to the Central Governorate, disadvantaging candidates with localized bases of support in and around Riffa. The districts of several current Islamist MPs, including that of the controversial Sh. Jasim al-Sa'idi, were even combined to force direct electoral face-offs among sitting Sunni legislators. Finally, while the Salafi and Muslim Brotherhood stronghold of Muharraq was spared redistricting, it was, on the other hand, the only governorate not to gain seats with the changes.

The state's clear aim, admitted even by loyalist groups supportive of the Crown Prince–backed electoral reforms,[19] is to preserve al-Wifaq's parliamentary minority while also averting the emergence of a populist, non-sectarian Sunni bloc along the lines of the Gathering of National Unity and to a lesser extent (for its apparent links to the Muslim Brotherhood) Sahwat al-Fatih. Having served its purpose of arresting the momentum of opposition demonstrations in 2011, Sunni nationalism is not a phenomenon the Bahraini state is pleased to see linger in the imagination of ordinary citizens, much less become institutionalized in the form of organized political societies in the first full elections since the uprising. Sunnis working together temporarily to block a Shi'a-led coup attempt is one thing—indeed, an act of loyalty to the ruling family—but Sunnis engaged in a sustained fight to secure a parliamentary majority over reliably pro-government tribal independents is a far more dubious project not to be taken passively by the state.

This conclusion speaks to a larger truth increasingly evident in Bahrain and across the Arab Gulf and Middle East region generally: growing domestic and international concern over the rise of transnational Sunni Islamist movements, and support for (to say nothing of direct involvement with) them among local populations. Once incited to political and even armed contestation by governments worried over domestic Shi'a activism and regional Iranian interference, now Sunni factions have arguably become a cure worse than the disease, threatening Gulf states' security and, with the growing notoriety of terrorist organizations such as the Islamic State (IS) group, their international reputations. Officials acknowledge that at least 100 Bahrainis have joined IS, including a former lieutenant in the Bahrain police force and the radical imam Turki al-Binali, "whose writings have set out the case for [IS leader] Abu Bakr al-Baghdadi's credentials as the righteous caliph to whom all Muslims owe allegiance."[20] Al-Binali was freely traveling and preaching in Bahrain at least until the end of 2013. In addition to urging fellow Sunnis to join the fight against Shi'a infidels in Syria and Iraq, Bahraini members of IS have in video messages implored their "'Sunni family' in Bahrain" to boycott the then-upcoming 2014 parliamentary elections, rejecting the legitimacy of the Al Khalifa regime.[21]

Thus, it is no longer just emboldened Shi'a populations that Bahrain and other Gulf governments must fear, and one observes that this post-Iraq War preoccupation has, in the wake of the Arab Spring, given way to a more general concern over religious-based movements of every stripe. The basic problem, as seen in the results of the Bahrain mass survey, is that such movements and individuals do not operate according to the prevailing economic-cum-political model of the Gulf: they simply do not play by the rules of the game. Driven neither by riches nor social status, religiously-motivated citizens can be neither purchased by governments nor easily deterred in their activism. Whereas in many parts of the world the single-minded pursuit of material gain is an unbecoming vice, in the Arab Gulf states the reverse case is true, a lack of avarice grounds for grave suspicion.

Finally, this shared perception of vulnerability has begotten a confluence of interests between the Gulf's anti-Islamist bloc on the one hand, especially Saudi Arabia and the United Arab Emirates with their near-fanatical hatred of the Muslim Brotherhood, and political-military patron the United States on the other. Beyond helping to re-cement relations strained by perceived American encouragement at the outset of the Arab uprisings, agreement over the danger posed by political Islam has lent newfound diplomatic leverage to the Gulf states, whose material and ideological support is seen as vital to combating IS and the regime of Bashar al-Assad in Syria. This has opened the door to domestically-advantageous issue linkages, namely moderation of the U.S. position on substantive political liberalization in the Gulf monarchies, including in Bahrain, the appetite for which in Washington has doubtless been dampened in any case by the seeming lack of any secular liberal alternative to traditional tribal rule. Of course, as seen in the preceding pages, such a lack is no accident.

Toward a New Arab Gulf Research Agenda

Such fundamental revisions in thinking about Arab Gulf politics call for corresponding changes to the present social science research agenda dominating studies of the region. The traditional focus on macro-level outcomes misses much of the politics in between, or rather gives the distinct impression of a want of political life altogether apart from top-down decisions of resource allocation made by calculating, interest-maximizing rulers whose only concern is the continued co-option of elite competitors via rent-funded patronage. Conspicuously absent from this view of the Arab Gulf states, then, are the vast majority of ordinary Gulf Arabs, an odd fact for a theory that purports to understand the bases of individual political behavior in resource-based regimes.

Not only has most quantitative research operated at the incorrect level of analysis, but, in so doing, it has employed an elastic notion of "rentierism" that has served to draw attention away from the Arab Gulf states as a particular class of regime. In seeking to find systemic relationships between country-level political outcomes and correlates of rentierism, such studies imply that every nation is to some degree a rentier state; that Luxembourg's democraticness owes in part to a lack of oil production. Yet the very data they employ suggest the opposite. Nearly all of the variation in country-level political outcomes attributed to oil in fact is attributable more simply to the distinct character of the Arab Gulf states, whose political institutions reflect a power structure little removed from the sheikhdoms of old. That none has become qualitatively more democratic is a counterfactual that in any case cannot be attributed to resource reliance. Were Saudi Arabia once democratic and then, following the discovery of oil, devolved into autocracy, that argument might be compelling. But demonstrating that a category of tribal monarchies remained tribal monarchies following some exogenous economic shock is not.

Dictating this research agenda in no small part has been a lack of requisite data, such collection impeded by a regional political environment generally hostile to public opinion research, and particularly hostile to survey research that would elucidate social or political conflict or sensitive demographics. Given this scarcity of individual-level data about the political views and activities of citizens in the region, to say nothing of their ethno-religious characteristics, it is little surprising that the behavioral assumptions of rentier theory have for so long escaped systematic empirical examination. At the same time, however, if the Bahrain survey could be undertaken in the midst of a security crackdown, then so too are other studies possible.

And such efforts are ongoing. The World Values Survey was administered for the first time in Qatar and Kuwait in 2010 and 2014, respectively, and the Arab Barometer for the first time in Saudi Arabia in 2011. The third wave of the Arab Barometer, scheduled for 2015, will be fielded in Qatar, Kuwait, and potentially one other GCC country. Yet, in order to expand on the findings of the present study, these two standard survey instruments must begin to include additional items, or to be more forceful in insisting that existing questions be fielded. Most significantly, neither the World Values Survey nor the Arab Barometer has succeeded in capturing the ethnic or confessional religious affiliation of Gulf Arab respondents, rendering these pioneering mass surveys unusable for purposes of elucidating group-based differences in political orientations and behaviors. Further, in light of the considerable sectarian interviewer effects observed in the Bahrain data, surveys in the region should begin to report basic demographic information about fieldworkers, including ascriptive group affiliation(s) in contexts where these are socially and politically salient, so that appropriate strategies may be adopted to avoid potentially substantial response bias.

Above all, having been now sufficiently reminded over the thirty-year ascendancy of rentier state theory about the importance of economic organization in determining the political character of Arab states, students of Middle East and Gulf politics should begin to proceed back in the other direction, to reevaluate the received stereotype of the economically and politically satiated oil sheikh in light of evolving domestic and regional conditions, including Shi'a populations increasingly insistent in their demands for political authority and influence, growing securitization of the region's government sectors, and the region's open political, ideological, and increasingly military competition between Saudi Arabia, Iran, and their respective allies. Such a reassessment of the continued efficacy of the rentier state paradigm demands, in the first place, a thorough interrogation of its conceptual underpinnings, which implies a return to the individual citizen as the primary unit of analysis, whether quantitative or qualitative. To understand the unforeseen longevity of the monarchies of the Arab Gulf, we must first understand the diverse processes and indeed active strategies by which they earn and preserve the political favor of ordinary Gulf Arabs.

Appendix

A Note on Interpretation

Reproduced in this Appendix are the model estimations underlying the empirical results presented in chapters 5 and 6. It is important to note that, although individual coefficient and corresponding significance estimates are reported here in the standard fashion, in many instances these are not directly interpretable owing to interactions between regressors. For multiplicative interaction terms, coefficient and standard error estimates do not represent unconditional marginal effects, but necessarily depend on the values of other variables included in the interaction(s).[1] For substantively meaningful interpretation of the regression results, then, including of marginal effects of independent variables of interest, see the relevant discussions in the text. The results here are presented for reference and to aid in statistical replication.

Models of Employment

Table Appendix 1. Models of employment sector and status (Figures 5.1–5.5)

	(1) Job Sector		(2) Job Sector (no police/military)		(3) Job Level	
Selection (Public Sector)						
Sect	2.244**	(0.001)	1.663+	(0.051)	−1.924**	(0.007)
Educ2	0.244	(0.541)	0.242	(0.548)	1.599*	(0.048)
Educ3	0.358	(0.400)	0.345	(0.427)	0.365	(0.632)
Educ4	0.363	(0.357)	0.331	(0.417)	−1.076	(0.186)
Educ2×Sect	−1.647*	(0.025)	−1.407	(0.117)	−0.574	(0.495)
Educ3×Sect	−2.179**	(0.006)	−1.724+	(0.063)	0.873	(0.272)
Educ4×Sect	−2.183**	(0.003)	−1.607+	(0.063)	2.243**	(0.004)
Female	0.487*	(0.014)	0.548**	(0.006)	−0.443	(0.109)
Age	−		−		0.0227+	(0.084)
λ (Mills)	−		−		0.214	(0.636)
Constant	−0.443	(0.245)	−0.447	(0.262)	3.760***	(0.001)
Behavioral (Working)						
Sect	0.0384	(0.802)	−0.0523	(0.740)	−0.0303	(0.850)
Educ2	0.244	(0.318)	0.277	(0.282)	0.275	(0.291)
Educ3	0.651*	(0.026)	0.743*	(0.014)	0.712*	(0.019)
Educ4	1.078***	(0.000)	1.188***	(0.000)	1.194***	(0.000)
Female	−0.127	(0.625)	−0.149	(0.572)	−0.160	(0.536)
Married	0.852***	(0.001)	0.885***	(0.000)	1.033***	(0.000)
Fem×Marr	−1.214***	(0.000)	−1.105***	(0.001)	−1.141***	(0.000)
Age	0.209**	(0.003)	0.195**	(0.007)	0.181*	(0.023)
Age^2	−0.00267**	(0.003)	−0.00249**	(0.007)	−0.00237*	(0.019)
Constant	−4.034***	(0.001)	−3.903**	(0.002)	−3.695**	(0.007)
N	402		390		84	

Notes: Probit models (1) and (2) estimated by heckprob in Stata; model (3) by heckman as described in the text; survey weights utilized; p-values in parentheses.
+ $p < 0.10$, * $p < 0.05$, ** $p < 0.01$, *** $p < 0.001$

Models of Political Attitudes and Behavior

Table Appendix 2. Models of political attitudes (Figures 5.7–5.8, 6.1)

	(1) Government Performance		(2) National Pride		(3) Legitimacy of 2006 Elections	
Sect	2.116	(0.286)	2.424$^+$	(0.097)	−0.332	(0.755)
Religiosity	−0.844**	(0.009)	−0.448*	(0.014)	−0.255	(0.122)
Relig × Sect	1.397**	(0.005)	−0.0473	(0.876)	0.293	(0.193)
Econ2	0.604$^+$	(0.054)	−0.279	(0.162)	−0.0160	(0.924)
Econ3	1.204*	(0.011)	−0.297	(0.297)	0.379	(0.156)
Econ2 × Sect	−0.562	(0.304)	−0.303	(0.412)	−0.0586	(0.828)
Econ3 × Sect	−1.182	(0.108)	−0.852	(0.140)	−0.390	(0.288)
Education	−0.0885	(0.483)	0.251**	(0.008)	−0.111	(0.103)
Educ × Sect	−0.133	(0.426)	−0.160	(0.277)	0.0710	(0.388)
Age	0.0293*	(0.021)	−0.00319	(0.648)	0.000163	(0.977)
Age × Sect	−0.0172	(0.329)	−0.0221$^+$	(0.085)	0.00558	(0.477)
Female	−0.153	(0.653)	0.414*	(0.020)	−0.119	(0.481)
Female × Sect	0.0643	(0.896)	−0.206	(0.508)	0.148	(0.509)
Diffsect	1.610***	(0.000)	0.311$^+$	(0.074)	1.035***	(0.000)
Diffsect × Sect	−1.512*	(0.014)	−0.446	(0.272)	−1.499***	(0.000)
Refuse	−3.459**	(0.007)	0.420	(0.498)	−0.696	(0.202)
Refuse × Sect	3.164$^+$	(0.075)	−1.187	(0.310)	1.096	(0.183)
Constant	5.593***	(0.000)			3.424***	(0.000)
N	373		383		340	
(Pseudo) R^2	0.545		0.1166		0.365	

Notes: Models (1) and (3) estimated by OLS; model (2) by ordered probit; survey weights utilized; *p*-values in parentheses.
$^+ p < 0.10, * p < 0.05, ** p < 0.01, *** p < 0.001$

Table Appendix 3. Models of political activities (Figures 5.9 and 5.10)

	(1) Attended Demonstration		(2) Petition / Meeting		(3) Voted in 2006 Elections	
Sect	1.388	(0.399)	2.245	(0.105)	−2.166⁺	(0.084)
Religiosity	0.362⁺	(0.085)	0.671**	(0.002)	0.648**	(0.002)
Sect × Relig	0.0657	(0.849)	−0.279	(0.399)	−0.207	(0.518)
Econ2	−0.158	(0.466)	−0.146	(0.503)	−0.0673	(0.753)
Econ3	0.0134	(0.969)	−0.00898	(0.979)	−0.0753	(0.823)
Econ2 × Sect	−0.457	(0.262)	0.0718	(0.860)	−0.338	(0.406)
Econ3 × Sect	−1.066⁺	(0.068)	0.237	(0.647)	−0.659	(0.213)
Education	0.149*	(0.024)	0.121*	(0.047)	0.108⁺	(0.084)
Age	−0.00850	(0.154)	−0.0137*	(0.020)	0.0310***	(0.000)
Female	−0.759***	(0.000)	−0.790***	(0.000)	−0.0676	(0.657)
Diffsect	0.228	(0.307)	0.590**	(0.009)	−0.0392	(0.848)
Diffsect × Sect	−0.356	(0.383)	−0.885*	(0.023)	0.377	(0.319)
Refuse	3.656***	(0.000)	3.220***	(0.001)	−0.751	(0.342)
Refuse × Sect	−1.951	(0.243)	−2.777*	(0.044)	3.313**	(0.007)
Constant	−3.775***	(0.001)	−3.412**	(0.002)	−1.028	(0.264)
N	382		388		382	
Pseudo R²	0.2421		0.2193		0.1283	

Notes: All models estimated by probit; survey weights utilized; *p*-values in parentheses.
⁺ $p < 0.10$, * $p < 0.05$, ** $p < 0.01$, *** $p < 0.001$

Table Appendix 4a. Models of political attitudes (Figures 5.4 and 5.11)

	(1) Policy Influence		(2) Economic Admin.		(3) Political Situation	
Sect	−0.801	(0.431)	−1.561*	(0.028)	−1.371	(0.229)
Religiosity	−0.0741	(0.534)	0.0474	(0.668)	−0.138	(0.196)
Relig × Sect	0.0916	(0.627)	0.104	(0.483)	0.170	(0.349)
Econ2	0.360*	(0.020)	0.161	(0.241)	0.167	(0.146)
Econ3	0.269	(0.112)	0.214	(0.223)	0.390+	(0.069)
Econ2 × Sect	0.317	(0.293)	0.448*	(0.034)	0.174	(0.418)
Econ3 × Sect	0.723*	(0.031)	0.811**	(0.001)	0.215	(0.483)
Education	−0.0408	(0.405)	0.0101	(0.829)	0.0366	(0.410)
Educ × Sect	0.00250	(0.973)	−0.0150	(0.787)	−0.191**	(0.002)
Age	0.000210	(0.962)	0.00152	(0.712)	0.00850*	(0.046)
Age × Sect	0.00182	(0.802)	0.00250	(0.633)	−0.00344	(0.583)
Female	−0.0501	(0.692)	−0.143	(0.206)	−0.168+	(0.086)
Female × Sect	−0.110	(0.563)	−0.0898	(0.554)	0.222	(0.170)
Diffsect	0.722***	(0.000)	0.313**	(0.005)	0.546***	(0.000)
Diffsect × Sect	−1.029***	(0.000)	−0.163	(0.327)	−0.487*	(0.046)
Refuse	−1.652***	(0.001)	−0.968*	(0.039)	−1.598***	(0.000)
Refuse × Sect	1.780*	(0.031)	1.841**	(0.001)	3.192***	(0.001)
Constant	3.690***	(0.000)	3.099***	(0.000)	2.816***	(0.000)
N	352		376		357	
R^2	0.401		0.260		0.387	

Notes: All models estimated by OLS; survey weights utilized; *p*-values in parentheses.

+ $p < 0.10$, * $p < 0.05$, ** $p < 0.01$, *** $p < 0.001$

Table Appendix 4b. Models of political attitudes (Table 5.4, continued)

	(4) Corruption		(5) Reform	
Sect	1.143	(0.134)	−0.631	(0.370)
Religiosity	0.00516	(0.961)	−0.125	(0.327)
Relig × Sect	−0.0168	(0.905)	0.0170	(0.917)
Econ2	−0.153	(0.143)	0.437**	(0.004)
Econ3	−0.0333	(0.826)	0.604***	(0.001)
Econ2 × Sect	−0.113	(0.541)	−0.0647	(0.778)
Econ3 × Sect	−0.292	(0.199)	−0.0592	(0.819)
Education	−0.00243	(0.956)	−0.103*	(0.045)
Educ × Sect	−0.0330	(0.573)	0.0579	(0.324)
Age	−0.00397	(0.288)	0.00185	(0.663)
Age × Sect	−0.00355	(0.515)	0.00387	(0.474)
Female	0.0993	(0.334)	0.217+	(0.055)
Female × Sect	0.335*	(0.020)	0.0382	(0.799)
Diffsect	−0.231*	(0.032)	0.169	(0.136)
Diffsect × Sect	0.252	(0.187)	−0.320+	(0.059)
Refuse	0.615	(0.149)	−0.479	(0.158)
Refuse × Sect	−1.344*	(0.025)	0.667	(0.199)
Constant	1.844***	(0.001)	3.698***	(0.000)
N	368		377	
R^2	0.154		0.199	

Notes: All models estimated by OLS; survey weights utilized; p-values in parentheses.
$^+ p < 0.10$, $^* p < 0.05$, $^{**} p < 0.01$, $^{***} p < 0.001$

Summary Statistics for Key Variables

Table Appendix 5. Summary of key variables

Variable	Sunni		Shi'i		
	Mean[a]	Std. Dev.	Mean	Std. Dev.	N
Sect	0.424	0.495	0.576	0.495	432
Religiosity	0.618	0.487	0.711	0.454	401
Econ1	0.129	0.336	0.242	0.429	429
Econ2	0.697	0.461	0.661	0.474	429
Econ3	0.174	0.380	0.0968	0.296	429
Education[b]	4.75	1.46	4.73	1.29	426
Educ1	0.159	0.367	0.133	0.340	426
Educ2	0.297	0.458	0.315	0.466	426
Educ3	0.143	0.351	0.166	0.372	426
Educ4	0.401	0.491	0.386	0.488	426
Age	37.1	14.50	35.1	13.6	429
Female	0.448	0.499	0.285	0.452	428
Diffsect	0.175	0.381	0.265	0.442	435
Refuse	0.930	0.129	0.920	0.144	435
Sector	0.505	0.502	0.385	0.488	249
Job level[c]	3.59	1.41	4.58	2.16	233
Working	0.568	0.497	0.612	0.488	249
Married	0.606	0.490	0.644	0.480	411

[a] Unweighted sample means.
[b] Original, continuous coding.
[c] Excluding police and military personnel.

Notes

Introduction

1. All production figures come from Arab Petroleum Research Center, *Arab Oil and Gas Directory* (Paris: Arab Petroleum Research Center, 2004).
2. Kingdom of Bahrain, *Census and Demographic Statistics*, Central Informatics Organization, 2001, p. 3. The latter proportion had been more or less stable since the first national census in 1941.
3. Mohammad Al Asoomi, "Oman and Bahrain Have Lot to Gain from GCC Plan," *Gulf News*, June 19, 2013.
4. Among other ways, such a diagnosis is disingenuous for overlooking a full-blown Shi'a-led uprising in Saudi Arabia's Eastern Province during the same period, which if it received less media attention lasted far longer and was far more widespread than protests in Oman. See, e.g., Toby Matthiesen, *Sectarian Gulf: Bahrain, Saudi Arabia, and the Arab Spring That Wasn't* (Palo Alto, CA: Stanford University Press, 2013).
5. *Bahrain-Saudi Arabia Boundary Agreement*, February 22, 1958.
6. Bernadette Michalski, *The Mineral Industry of Bahrain*, United States Geological Survey, 1996, p. 1.
7. Kingdom of Bahrain, *Annual Economic Review*, Economic Development Board, 2010, p. 43.
8. Anon., "GCC's Oil and Gas Annual Earnings Hit $US737.5 Billion," *Oil Review Middle East*, March 19, 2013.
9. See note 4.
10. Fahim I. Qubain, "Social Classes and Tensions in Bahrain," *Middle East Journal* 9.3 (1955): 269.
11. While it is still too early to close the book on the February 14th uprising, useful overviews include, in chronological order, International Crisis Group, "Popular Protest in North Africa and the Middle East (VIII): Bahrain's Rocky Road to Reform," *Middle East Report No. 111*, July 28, 2011; Jane Kinninmont, "Bahrain: Beyond the Impasse" (London: The Royal Institute for International Affairs, 2012); and Kenneth Katzman, "Bahrain: Reform, Security, and U.S. Policy," United States Congressional Research Service, April 1, 2013. For a far more extensive account of the initial period of unrest and the government's response, see the 513-page final report of the Bahrain Independent Commission of Inquiry. The independent panel of foreign legal experts was established in July 2011 by royal order of King Hamad, tasked with investigating the uprising and its aftermath in the face of claims of widespread human rights abuses. Bahrain Independent Commission of Inquiry, *Report of the Bahrain Independent Commission of Inquiry*, November 23, 2011 (rev. December 10), available at www.bici.org.bh/BICIreport EN.pdf.
12. Some even speak of a "new Middle East Cold War," as coined in Bill Spindle and Margaret Coker, "The New Cold War," *Wall Street Journal*, April 16, 2011. Oman, which is neither Sunni-led nor actively involved in the conflict, is an exception.
13. Qubain, "Social Classes and Tension in Bahrain."

Chapter 1

1. Hossein Mahdavy, "Patterns and Problems of Economic Development in Rentier States: The Case of Iran," in M. A. Cook, ed., *Studies in the Economic History of the Middle East: From the Rise of Islam to the Present Day* (London: Oxford University Press, 1970), p. 428; Hazem Beblawi, "The Rentier State in the Arab World," in Hazem Beblawi and Giacomo Luciani, eds., *The Rentier State: Nation, State and Integration in the Arab World*, Vol. 2 (London: Croon Helm, 1987), p. 51.

2. Gary G. Sick, "The Coming Crisis in the Persian Gulf," in Gary G. Sick and Lawrence G. Potter, eds., *The Persian Gulf at the Millennium: Essays on Politics, Economy, Security, and Religion* (New York: St. Martin's, 1997), p. 11.

3. Dirk Vandewalle, "Political Aspects of State Building in Rentier Economies: Algeria and Libya Compared," in Beblawi and Luciani, eds., *The Rentier State*, p. 160.

4. Giacomo Luciani, "Allocation vs. Production States: A Theoretical Framework," in Beblawi and Luciani, eds., *The Rentier State*, p. 10.

5. Hazem Beblawi, "The Rentier State in the Arab World," in Giacomo Luciani, ed., *The Arab State* (London: Routledge, 1990), p. 91.

6. Mohammed Zaher, "GCC: Fiscal Stimulus and Reforms Are Optimal Choice under Current Circumstances," *GCC Research Note*, NBK (National Bank of Kuwait), April 2, 2009, available at www.kuwait.nbk.com.

7. Ibid.

8. Cited in Suliman Al-Atiqi, "Laboring against Themselves," Sada blog, Carnegie Endowment for International Peace, February 26, 2013.

9. F. Gregory Gause, "Regional Influences on Experiments in Political Liberalization in the Arab World," in Rex Brynen, Bahgat Korany, and Paul Noble, eds., *Political Liberalization and Democratization in the Arab World*, Vol. 1, *Theoretical Perspectives* (Boulder, CO: Lynne Rienner, 1995), p. 77.

10. Nazih Ayubi, "Arab Bureaucracies: Expanding Size, Changing Roles," in Luciani, ed., *The Arab State*, p. 144.

11. Some theorists have expanded this list of relevant variables to include, in particular, repression of political opponents by rent-funded security forces. [E.g., Theda Skocpol, "Rentier State and Shi'a Islam in the Iranian Revolution," *Theory and Society* 11.3 (1982): 265–283.] Ross also tests the argument that resource-based wealth does not lead to democracy because it does not involve the typical changes in sociocultural attitudes that come via the typical process of modernization. ["Does Oil Hinder Democracy?" *World Politics* 53.3 (2001): 325–361.] Such additions, however, are less convincing than the original mechanisms described by Beblawi and Luciani, a fact that Ross himself now concedes, admitting, "I no longer find support for two of the three mechanisms discussed in Ross [2001]; nor is there evidence to support mechanisms alleged by others. The only mechanism that seems to matter is the *rentier* effect": that is, "the combination of low taxes and high government spending that seems to dampen support for democratic transitions" ["Oil and Democracy Revisited," unpublished manuscript, www.sscnet .ucla.edu/polisci/faculty/ross/Oil and Democracy Revisited.pdf., March 2, 2009, pp. 2, 25].

12. E.g., Jocelyn Mitchell, "Beyond Allocation: The Politics of Legitimacy in Qatar," PhD diss., Georgetown University, Washington, DC, 2013. Another partial exception is Sean Foley, *The Arab Gulf States: Beyond Oil and Islam* (Boulder, CO: Lynne Rienner, 2010).

13. The most developed such theoretical update is Matthew Gray, "A Theory of 'Late Rentierism' in the Arab States of the Gulf," *Occasional Paper No. 7*, Center for International and Regional Studies, Georgetown School of Foreign Service in Doha, 2011.

14. This popular variable from the Polity IV dataset places states on a scale from full autocracy (−10) to full democracy (10) for the years 1800–2012. Data available at www.systemicpeace .org/polity/polity4.htm. See, e.g., Keith Jaggers and Ted Robert Gurr, "Tracking Democracy's Third Wave with the Polity III Data," *Journal of Peace Research* 32.4 (1995): 469–482.

15. Inexplicably, Bahrain's score temporarily dipped to −5 in 2010, before reverting to −8 the next year and then to −10 in 2012.

16. Michael L. Ross, "Oil, Islam, and Women," *American Political Science Review* 102.1 (2008): 111–112. The paper, incidentally, won the 2009 award for best article in the *American Political Science Review.*

17. Data obtained from the author for purposes of replication.

18. If one would instead omit only the six Gulf Cooperation Council states, the statistical significance of the bivariate relationship drops from 0.0000 to 0.0413, and the adjusted R-squared from 0.10 to 0.02. Groh and Rothschild undertake a similar procedure, excluding the six Arab Gulf states along with Yemen, in "Oil, Islam, Women, and Geography: A Comment on Ross (2008)," *Quarterly Journal of Political Science* 7.1 (2012): 69–87.

19. Luciani, "Allocation vs. Production States," in Beblawi and Luciani, eds., *The Rentier State*, p. 63.

20. Gwenn Okruhlik, "Rentier Wealth, Unruly Law, and the Rise of Opposition: The Political Economy of Oil States," *Comparative Politics* 31.3 (1999): 297.

21. A possible third explanation, based on the so-called repression effect suggested by Skocpol and tested by Ross (cf. note 11), is that states' security forces are for some reason unable to carry out their normal, rent-funded duty of suppressing government opposition. That such a breakdown has occurred, however, is theoretically posterior to and indeed only begs the question of *why* it has occurred. In which case we are returned once again to the same two possibilities: structural difficulties (perhaps the state no longer has the money to fund the secret police; they lack the manpower to keep up with a growing opposition; the police are growing too powerful and leaders wish to reign them in; etc.); or a mischaracterization of the nature of the opposition (i.e., government critics understand the consequences of acting out but choose to do so anyway).

22. The short-lived demonstrations witnessed in Oman in 2011 are arguably such a case. See James Worrall, "Oman: The 'Forgotten' Corner of the Arab Spring," *Middle East Policy* 19.3 (2012): 98–115; and Marc Valeri, "The Qaboos-State under the test of the 'Omani Spring': Are the Regime's Answers Up to Expectations?" *Les Dossiers du CERI* (Paris: Sciences Po, 2011); "Identity Politics and Nation-Building under Sultan Qaboos," in Larry Potter, ed., *Sectarian Politics in the Persian Gulf* (London/New York: Hurst/Oxford University Press, 2013).

23. E.g., Kristin Smith Diwan, "Kuwait's Constitutional Showdown," *Foreign Policy*, November 17, 2011; and, "Kuwait's Balancing Act," *Foreign Policy*, October 23, 2012.

24. Kristian Coates Ulrichsen, "The UAE: Holding Back the Tide," *Open Democracy*, August 5, 2012. See also Christopher M. Davidson, "The United Arab Emirates: Frontiers of the Arab Spring," *Open Democracy*, September 8, 2012.

25. For more on this topic, see Justin Gengler, "Understanding Sectarianism in the Persian Gulf," in Larry Potter, ed., *Sectarian Politics in the Persian Gulf.*

26. Daniel M. Corstange, "Institutions and Ethnic Politics in Lebanon and Yemen," PhD diss., University of Michigan, 2008, pp. 132–135.

27. Here as elsewhere I speak only of Gulf nationals and exclude any systematic consideration of expatriate laborers, which is a subject demanding a separate inquiry unto itself.

28. Additionally, one might argue that in those Gulf states in which the primary social division is between citizens and non-citizens, such as Qatar and the United Arab Emirates, this dichotomy effectively takes on the same political role as that of other ascriptive social distinctions,

begetting a sort of group politics that is not qualitatively different from that witnessed elsewhere in the region. Indeed, I have even heard some Gulf nationals observe that, whatever its other drawbacks, the extreme citizen-expatriate imbalance in parts of the region is ultimately a political boon for governments, as it helps to coalesce a sense of national identity, even if a negative one: "I am *not* an expatriate" [Personal correspondence, Kuwait, May 2013].

For recent extended treatments of state manipulation of (sectarian) group identities in the Gulf context, see Matthiesen, *Sectarian Gulf*; Frederic Wehrey, *Sectarian Politics in the Gulf: From the Iraq War to the Arab Uprisings* (New York: Columbia University Press, 2012); and Potter, ed., *Sectarian Politics in the Persian Gulf*.

29. Douglas A. Yates, *The Rentier State in Africa: Oil Rent Dependency and Neocolonialism in the Republic of Gabon* (Trenton, NJ: Africa World Press, 1996), p. 35.

30. Luciani, "Allocation vs. Production States," in Beblawi and Luciani, eds., *The Rentier State*, p. 74.

31. Ibid.

32. Ibid., p. 76.

33. Not to say political "parties." Bahrain allows and regulates the existence of political "societies," while Kuwait permits political "blocs." In practice, however, the difference is largely semantic.

34. Anon., "Kuwait MPs Expelled for Mourning Mughniyah," *Al-Arabiya*, February 20, 2008.

35. For an overview of Kuwait's parliamentary landscape, see Kenneth Katzman, "Kuwait: Security, Reform, and U.S. Policy," United States Congressional Research Service, March 29, 2013, p. 9.

36. Karim Hamad, "بعد إعلان النتائج النهائية: سيطرة دينية على المجلس," *Akhbar Al-Khaleej*, December 4, 2006.

37. Most obvious in this respect is the secular-leftist National Democratic Action Society, whose secretary general was among the first of the opposition leaders arrested in March 2011. More generally, the group's cross-sectarian membership and following made it a target of electoral machinations throughout the 2000s, during which time it failed despite considerable popularity to capture even a single seat in parliament. See, e.g., Justin Gengler, "Bahrain's Sunni Awakening," *Middle East Research and Information Project*, January 17, 2013; "Bahrain: A Special Case," in Fatima Ayub, ed., *What Does the Gulf Think about the Arab Awakening?* (London: European Council on Foreign Relations, 2013), pp. 16–18; and pp. 82 and 143 in this book.

38. Most notably, the al-Haqq Movement for Liberty and Democracy (al-Haqq) and the Islamic Fidelity Movement (al-Wafa'). See chap. 3, pp. 77ff.

39. Ben Birnbaum, "Pro-government Cleric to Start Own Party in Bahrain," *Washington Times*, August 9, 2011.

40. Kanchan Chandra, *Why Ethnic Parties Succeed* (Cambridge, UK: Cambridge University Press, 2004).

41. Benedict Anderson, *Imagined Communities: Reflections on the Origin and Spread of Nationalism* (London: Verso, 1983).

42. Donald L. Horowitz, *Ethnic Groups in Conflict* (Berkeley: University of California Press, 1985).

43. See Gengler, "Understanding Sectarianism in the Persian Gulf," for a complete version of this argument in the Gulf context.

44. Corstange, "Institutions and Ethnic Politics in Lebanon and Yemen," p. 20.

45. *'ajam* (sing. *'ijmī*): Literally, one who is illiterate in language; silent; mute. Though the term can refer to non-Arabic-speaking peoples more generally and is sometimes considered an ethnic slur, in Bahrain it is used widely to denote Shi'a of Persian origin, who have a neighborhood in Manama (*fareej al-'ajam*) as well as several prominent religious institutions named

for them. Many 'Ajam descend from those originally recruited for work in the nascent oil sector, in which Persians made up a large segment of the labor force. Some eventually were granted Bahraini citizenship (cf. chap. 2, note 33), while others remain stateless. Louër tells that those who did receive citizenship "were generally well-connected to the affluent Iranian merchant families close to the ruling dynasty." [Laurence Louër, "The Political Impact of Labor Migration in Bahrain," *City & Society* 20.1 (2008): 52.]

46. *huwala* (sing. *huwalī*): Sunnis who migrated from the Iranian side of the Gulf littoral in the late nineteenth and first half of the twentieth centuries.

Indeed, far from common knowledge, agreement does not exist even over the reputed descent of the Huwala. Louër writes, "They claim ancient Arab tribal origin, which the Baharna usually deny by saying they are Iranians who try to pass as Arabs to gain the favor of the Al-Khalifa. Many Huwala are part of the merchant oligarchy intimately tied to the rulers. They are overrepresented in the directorial positions of oil and aluminium companies." [Louër, "The Political Impact of Labor Migration in Bahrain," p. 39.]

47. Melissa S. Williams, *Voice, Trust, and Memory: Marginalized Groups and the Failings of Liberal Representation* (Princeton, NJ: Princeton University Press, 2010), pp. 108ff.

48. Of course, some may also harbor anti-Shiʻa or anti-Sunni orientations independent of domestic political considerations, for example on doctrinal grounds.

49. Bruce Bueno de Mesquita, James D. Morrow, Randolph M. Siverson, and Alastair Smith, *The Logic of Political Survival* (Cambridge, MA: MIT Press, 2003).

50. This latter question of gaining and consolidating family support through economic largesse is largely ignored in the rentier state literature, wherein the citizen-state relationship is paramount. Yet, particularly in instances of succession or potentially painful political transformation, one observes no little resources spent attempting to secure the support of the ruling family itself. Such was the case, for instance, following the 1999 succession of Bahrain's King Hamad upon the sudden death of his father. The new king's munificence was soon spread among royals and ordinary citizens alike. Citizens saw housing grants and loan forgiveness amounting to more than half a billion dollars, while members of the Al Khalifa saw increases in their monthly stipends as well as their representation in senior government positions. See, e.g., Abdulhadi Khalaf, "Al Khalifa, Hamad bin Isa (1950–)," in *The Biographical Encyclopedia of the Modern Middle East and North Africa* (Farmington Hills, MI: Thomson Gale, 2007); and Justin Gengler, 2013, "Royal Factionalism, the Khawalid, and the Securitization of the 'Shiʻa Problem' in Bahrain," *Journal of Arabian Studies* 3.1 (2013): 53–79.

51. Of course, population size and makeup may be altered to suit a particular strategy, most often political segmentation. Cf. chap. 6, pp. 154f.

52. Justin Gengler, "How Radical Are Bahrain's Shia?" *Foreign Affairs*, May 15, 2011.

53. Several of these initial efforts are summarized in Thomas Fuller, "Bahrain's Promised Spending Fails to Quell Dissent," *New York Times*, March 6, 2011.

54. The latter, announced in November 2012, will include "the world's biggest shopping mall, a Universal family theme park and a park that is a third bigger than Hyde Park in London." See, e.g., Andy Sambidge, "Dubai Ruler Announces New Mega City Project," *Arabian Business*, November 24, 2012.

55. E.g., Daniel Brumberg, "Transforming the Arab World's Protection-Racket Politics," Journal of Democracy 24.3 (2013): 88–103. On Bahrain and Saudi Arabia, see Gengler, "Bahrain's Sunni Awakening"; and Toby Matthiesen, "Saudi Arabia's Shiite Escalation," *Foreign Policy*, July 10, 2012.

56. The 1783 conquest of the island by the ruling Al Khalifa family was assisted by numerous allied tribes, most of which were rewarded with gifts of land and control over feudal estates. This close political alliance has persisted to the present. For a possible explanation of this

shift in state strategy beginning in 1999, see Gengler, "Royal Factionalism, the Khawalid, and the Securitization of 'the Shīʿa Problem' in Bahrain."

57. This range represents the 95 percent confidence internal for a mean estimate of 57.6 percent, which is the percentage of Shiʿa respondents in the Bahrain survey. Cf. chap. 4, p. 96.

58. Quoted in Anon., "Bahrain's Pre-election Jitters," *The Economist*, October 14, 2010.

59. Gengler, "Understanding Sectarianism in the Persian Gulf," pp. 34–47.

60. Though it is beyond the scope of the present study, one might also include a third conflict that continues to shape the direction of Bahraini politics, namely the one fought between senior Al Khalifa themselves. On this see Gengler, "Royal Factionalism, the Khawalid, and the Securitization of 'the Shīʿa Problem' in Bahrain."

Chapter 2

1. Since the 1941 census, Sunnis and Shiʿis have been categorized simply as "Muslims." A more reliable estimate of the country's present sectarian balance follows in chapter 4.

2. If one should wonder about its sectarian overtones, "Fath al-Islam" was also, for instance, the name adopted by Sunni militants who made headlines in summer 2007 for an armed rebellion inside a Palestinian refugee camp in Lebanon that had the larger aim of striking at Hizballah, arousing tensions with the latter and its Shiʿa supporters. See Robert Worth and Nada Bakri, "Hezballah Ignites a Sectarian Fuse in Lebanon," *New York Times*, May 18, 2008.

3. Personal interview, Bahrain, May 2009. Not coincidentally, Al-Fatih soon emerged as the de facto symbol of the Sunni-led pro-government movement sparked by the February 2011 uprising. At the height of the crisis in February and March 2011, the mosque itself served as the base of counter-revolutionary protests by pro-regime Sunnis. This largely spontaneous response later coalesced into several new populist Sunni political factions, prominent among them Sahwat al-Fatih ("the Al-Fatih Awakening") and the Al-Fatih Youth Coalition. There even emerged an Al-Fatih Group for Electronic Jihad, meant to combat an effective international media campaign waged by opposition activists.

4. Fuad Khuri, *Tribe and State in Bahrain* (Chicago: University of Chicago Press, 1980), p. 28.

5. Laurence Louër, *Transnational Shia Politics* (New York: Columbia University Press, 2008), p. 23.

6. Like the myth of Ancient Bahrain, the term *"baḥārna"* is also popular, Louër tells [ibid., p. 12], among Saudi Shiʿa, especially among intellectuals and political activists, who use it to denote Shiʿa living in the Eastern Province. I have heard it used widely in Qatar also as a generic name for Qatari and Gulf Arab Shiʿa.

7. In Charles Belgrave, *The Pirate Coast* (Beirut: Librairie du Liban, 1960); and *Personal Column* (Beirut: Librairie du Liban, 1972).

8. See Bahrain Centre for Human Rights (BCHR), "Banning one of the Most Significant Historic Books in the History of Bahrain," May 25, 2010, available at www.bahrainrights.org/en/node/3105.

9. See "Papers of Charles Dalrymple-Belgrave, 1926–1957," available at www.scribd.com/doc/16225787.

10. Mai bint Muhammad Al Khalifa, *From the Surroundings of Kufa to Bahrain: The Carmathian, from an Idea to a State* (Beirut: Arab Institute for Studies and Publishing, 1999); and *Charles Belgrave: Biography and Diary, 1926–1957* (Beirut: Arab Institute for Studies and Publishing, 2000).

11. Clive Holes, "Dialect and National Identity: The Cultural Politics of Self-Representation in Bahraini *Musalsalāt*," in Paul Dresch and James Piscatori, eds., *Monarchies and Nations: Globalization and Identity in the Arab States of the Gulf* (London: I. B. Tauris, 2005).

12. The Shiʻa uprising of 1994–1999 is covered in detail in, e.g., Louay Bahry, "The Opposition in Bahrain: A Bellwether for the Gulf?" *Middle East Policy* 5.2 (1997): 42–57; Munira Fakhro, "The Uprising in Bahrain: An Assessment," in Gary Sick and Lawrence Potter, eds., *The Persian Gulf at the Millennium* (New York: St. Martin's, 1997), pp. 167–188; Adeed Darwish, "Rebellion in Bahrain," *Middle East Review of International Affairs* 3.1 (1999): 84–87; Louay Bahry, "The Socio-economic Foundations of the Shiite Opposition in Bahrain," *Mediterranean Quarterly* 11.3 (2000): 129–143; Fred H. Lawson, "Repertoires of Contention in Contemporary Bahrain," in Quintan Wiktorowicz, ed., *Islamic Activism: A Social Movement Theory Approach* (Bloomington: Indiana University Press, 2004); and J. E. Peterson, "Bahrain: The 1994–1999 Uprising," *Arabian Peninsula Background Note, No. APBN-002*, 2004, available at www.JEPeterson.net.

13. For more on King Hamad's reform initiative, see, e.g., Abdulhadi Khalaf, "The New Amir of Bahrain: Marching Sideways," *Civil Society* 9.100 (2000): 6–13; J. E. Peterson, "Bahrain's First Reforms under Amir Hamad," *Asian Affairs* 33.2 (2002): 216–227; Edward Burke, "Bahrain: Reaching a Threshold," Working paper presented at El Fundación para las Relaciones Internacionales y el Diálogo Exterior (FRIDE), Madrid, June 5, 2008, available at www.fride.org/pub lication/452/bahrain-reaching-a-threshold.html; and Abdulhadi Khalaf, "The Outcome of a Ten-Year Process of Political Reform in Bahrain," *Arab Reform Brief No. 24*, 2008, available at www .arab-reform.net/sites/default/files/ARB.23_Abdulhadi_Khalaf_ENG.pdf.

14. J. E. Peterson, "The Promise and Reality of Bahraini Reforms," in Joshua Teitelbaum, ed., *Political Liberalization in the Gulf* (New York: Columbia University Press, 2008). In fact, the event remains so infamous that one can still find a video of the entire ceremony, including Shaikh Hamad's signature, online. Since its posting on YouTube in August 2007, it has been viewed almost 300,000 times. See (in Arabic) "ملك البحرين يحلف على القرآن وينكث" ["The King of Bahrain Swears on the Qur'an and [Then] Reneges"], available at www.youtube.com/watch ?v=-ux3dIonYpQ.

15. Article 120, paragraph C of the February 14, 2002, Constitution of the Kingdom of Bahrain. Quoted in Steven M. Wright, "Fixing the Kingdom: Political Evolution and Socio-economic Challenges in Bahrain" (Doha: Center for International and Regional Studies, 2008). See Wright's work for a thorough analysis of the various provisions of the 2002 Constitution.

16. Khalaf, "The Outcome of a Ten-Year Process of Political Reform in Bahrain," pp. 4–6.

17. Wright, "Fixing the Kingdom." These electoral constituencies were once again unilaterally redrawn in September 2014. See "Royal decree demarcates electoral districts, constituencies and electoral subcommittees," Information Affairs Authority, Kingdom of Bahrain, September 23, 2014, available at http://www.iaa.bh/news-details.aspx?id=463.

18. Ebrahim Sharif, personal interview, Bahrain, May 11, 2009.

19. Wright, "Fixing the Kingdom," p. 6.

20. Quoted in Mansour al-Jamri, "State and Civil Society in Bahrain," paper presented at the Annual Conference of the Middle East Studies Association, Chicago, December 9, 1998.

21. The video and an English transcript can be found at Bahrain Centre for Human Rights, "Documentary Film Script: The Political Naturalization in Bahrain," 2002, available at www .bahrainrights.org/node/269.

22. See Habib Trabelsi, "Bahrain's Shiite Muslims Cry Foul over Dual Nationality Plan," *Khaleej Times,* June 16, 2002.

23. See HAQ: Movement of Liberties and Democracy—Bahrain, "Motivated Change of Demography. Infringements of Political Rights and Inadequate Living Standards," report submitted

to the Universal Periodic Review Working Group of the United Nations Office of the High Commissioner for Human Rights, November 19, 2007, available at http://lib.ohchr.org/HR-Bodies/UPR/Documents/Session1/BH/MLD_BHR_UPR_S1_2008_Movement ofLibertiesand-DemocracyHAQ_%20uprsubmission.pdf.

24. Salah al-Bandar, "البحرين: الخيار الديموقرطي وآليات الإقصاء" ["Bahrain: The Choice of Democracy and the Machinery of Exclusion"], unpublished report prepared by the Gulf Centre for Democratic Development, 2006, available (in Arabic) at www.bahrainrights.org/files/albandar .pdf. The so-called Bandargate scandal received no little press and is treated in, e.g., Burke, "Bahrain: Reaching a Threshold"; Louër, *Transnational Shia Politics*; and Wright, "Fixing the Kingdom."

25. Ahmad bin ʻAtiyatallah is also a nephew of Royal Court Minister Khalid bin Ahmad. See Justin Gengler, "Royal Factionalism, the Khawalid, and the Securitization of the 'Shiʻa Problem' in Bahrain," *Journal of Arabian Studies* 3.1 (2013): 53–79.

26. "تصور للنهوض بالوضع العام للطائفة السنية في مملكة البحرين" ["A Proposal to Promote the General Situation of the Sunni Sect in the Kingdom of Bahrain"], unpublished paper dated September 1, 2005, in Salah al-Bandar, "البحرين: الخيار الديموقرطي وآليات الإقصاء," pp. 184–202.

27. All quotations translated from ibid., pp. 185–186. Emphasis added.

28. Translated from ibid., p. 186.

29. For still more on al-Bandar's report, see Bahrain Centre for Human Rights, "The Al Bander Report: What It Says and What It Means," 2006, available at www.bahrainrights.org /node/528.

30. Jane Kinninmont, "Bahrain: Beyond the Impasse: Programme Report" (London: Royal Institute of International Affairs, 2012), p. 5.

31. Anon., "BANDARGATE!" *Gulf Daily News*, September 24, 2006. See also the in-depth follow-up, Anon., "BANDARGATE: The Unanswered Questions," *Gulf Daily News*, September 27, 2006.

32. A full English translation of the letter can be found at Bahrain Centre for Human Rights, "A Petition from a Hundred Prominent Figures and Activists to the King of Bahrain," dated October 13, 2006, available at www.bahrainrights.org/node/610.

33. According to Ebrahim Sharif, the Sunni head of the leftist National Democratic Action Society (Waʻad), the published figures indicated that around 60,000 people had been naturalized since 2001. This was based on the average population growth rate for the preceding years, which was around 2.4 percent. As the new data implied a growth rate of about 4.2 percent from 2001 to 2007, they suggested an annual naturalization rate of approximately 1.8 percent, or about 9,000 citizens per year. All of whom are assumed to be Sunnis, as no Shiʻa are known to have been naturalized since several thousand second- and third-generation stateless individuals of Persian origin were granted citizenship in 2001 as part of Hamad's reforms. Personal interview, Bahrain, May 2009.

34. Quoted in Muhammad al-Aʻali, "Session Disrupted over 'Bandargate,'" *Gulf Daily News*, March 12, 2008.

35. Quoted in Robin Wright and Peter Baker, "Iraq, Jordan See Threat to Elections from Iran," *Washington Post*, December 8, 2004.

36. Louër, *Transnational Shia Politics*, p. 244.

37. Lit., "a council"; a regular public audience held by a notable.

38. Quoted in Habib Toumi, "Call to Lift Bahrain Parliamentary Immunity over Offensive Remarks," *Gulf News*, February 26, 2009.

39. As implied by the name, the law applies only to Sunni Muslims. Shiʻa religious leaders, several of whom then sat in parliament, were able to resist the promulgation of a correspond-

ing Shi'a Family Law. For background and analysis, see Jane Kinninmont, "Framing the Family Law: A Case Study of Bahrain's Identity Politics," *Journal of Arabian Studies* 1.1 (2011): 53–68.

40. Sing. *ma'tam*: lit., "funeral house"; known elsewhere as a *ḥusayniyya*, in Bahrain the *ma'tam* is a Shi'a center of learning and of worship on special religious occasions and holidays, especially during the month of Muharram. Most *mawātim* are copiously decorated with images of the Imam Husain; popular *marāji'* such as Ayatollahs al-Sistani, Khomeini, and Khamene'i; and large banners embroidered with Qur'anic verses and exaltations.

41. Cf. conclusion of note 33.

42. A reference to a failed plot carried out in December 1981 by the Islamic Front for the Liberation of Bahrain, allegedly with support from Iran. For a recent revisiting of this episode, see Hasan Tariq Alhasan, "The Role of Iran in the Failed Coup of 1981: The IFLB in Bahrain," *Middle East Journal* 65.4 (2011): 603–617. Of course, most Sunnis would also describe the February 14th uprising as no less of an Iranian-backed coup attempt.

43. Personal interview, Bahrain, May 2009.

44. This and the following quotation are from a personal interview, April 2009.

45. The *khums*, literally "one-fifth," in fact applies only to a family's yearly surplus income or to a windfall gain. This Shi'a-specific tax, considered an unlawful "innovation" by Salafis and other Sunnis, typically is paid to a particular *mujtahid* or *marja'*, though of course not all of these will be "mullahs in Iran."

46. *Zakat*, or alms-giving to the poor, is common to Sunnis and Shi'a and is one of the five Sunni "pillars of Islam."

47. In September 2014, the Huthis accomplished the highly unlikely feat of capturing the capital Sana'a and taking control of most government ministries.

48. Salman al-Dawsari, "البحرين: نائب رئيس اللجنة التشريعية في البرلمان يتهم الوفاق بـ 'تحركات مشبوهة'مع الحوثيين في اليمن" ["Bahrain: Vice-Chairman of the Legislative Committee in Parliament Accuses the al-Wifaq Opposition of 'Suspicious Dealings' with the Huthis of Yemen"], *al-Sharq al-Awsat*, August 24, 2009.

49. Quoted in ibid.

50. And, in fact, the Saudi Isma'ilis have often been accused by the Yemeni government of just that. See, e.g., Anon., "شيعة السعودية يطالبون بحقوق أفضل في الجنوب: واليمن يتهم إيران بمحاولة إقامة دولة 'شيعية'" ["The Shi'a of Saudi Arabia Call for Better Rights in the South; and Yemen Accuses Iran of Trying to Create a Shi'a Mini-State"], *Ma'rib Press*, May 27, 2009, available at http://marebpress.net/news_details.php?lng=arabic&sid=16782.

51. Ali Abdullah Saleh, "في حديثه لـ 'واجه الصحافة'مع داود الشريان، الرئيس اليمني: لا وجود أمريكياً في أراضينا ولن أترشح للرئاسة" ["In His Interview with *Meet the Press* with Dawud al-Sharyan, the Yemeni President: 'There is not a single American on our soil, and I won't run for the presidency,'" televised interview with *Al-Arabiya*, March 19, 2010. The video and a partial transcript are available at www.alarabiya.net/articles/2010/03/19/103454.html. The quotation above corresponds to around 9:50 to 11:00.

52. From ibid., at approximately 14:30 to 15:00.

53. London in particular is known as "a foremost centre of Shia activity" worldwide. It was here, for example, that the Islamic Bahrain Freedom Movement (IBFM) was founded in 1982. An offshoot of the Bahraini branch of Iraq's Shi'a al-Da'wa party, itself active in London following crackdowns in Iraq and the Gulf throughout the 1970s and 1980s, the IBFM continues to be a thorn in the side of the Bahraini regime, maintaining a popular bilingual website (www.vob.org) and electronic newsletter called "Voice of Bahrain" that catalogue Al Khalifa abuses. The group's real success, however, lies in its effective targeting of English-speaking audiences, which it does by lobbying individual politicians, organizing parliamentary and congressional

hearings on Bahrain, and working with human rights bodies like Amnesty International and the UN. See Louër, *Transnational Shia Politics*, pp. 202–203, 266.

54. For reasons that are not clear to me, neither the president nor ordinary Yemeni Zaydis would consider themselves "Shiʿa," a label they reserve for Twelver Shiʿa, a fact that has occasioned many an argument with Yemeni friends.

55. Louër, *Transnational Shia Politics*, pp. 45–63.

56. Ibid., p. 258.

57. Ibid., p. 245.

58. This is not to say that both traditions are represented in all areas. In addition to sovereign ministries as well as the police and armed forces, disproportionately few Shiʿa are employed in the Ministry of Education, for instance, which continues to reject calls to include Jaʿafari perspectives in the Maliki-dominated school curriculum.

59. Ibid., p. 255.

60. Horowitz, *Ethnic Groups in Conflict*, p. 187.

61. Military reliance upon non-aligned foreigners has a long history in the Arab world, but in modern times it is a particularly salient feature of the Arab Gulf region. Khuri [1980, 51 ff.] documents the use in Bahrain of so-called *banī khdayr* ("the green stock")—Sunnis with "no clear tribal origin: Baluchis, Omanis, 'stray' Arabs who lost tribal affiliation, and people of African origin"—since the time after the Al Khalifa arrival.

62. Quoted in International Crisis Group, "Bahrain's Sectarian Challenge," *Middle East Report No. 40*, May 6, 2005, p. 8.

63. The individual accused of leaking this list of names, himself an employee of an unspecified ministry, was imprisoned and purportedly "offered . . . a bargain in return for his release, on the condition that he signs a statement in which he accuses both Nabeel Rajab—President of the BCHR—and women activist Layla Dishti—administrator of www.bahrainonline.org [a popular Shiʿa opposition web forum where the list of names first appeared]—that they incited and funded him to publish those names" [Bahrain Centre for Human Rights, "Arbitrary Detention of a Citizen for Disseminating Information on the National Security Apparatus," June 21, 2009, available at www.bahrainrights.org/en/node/2914]. That the authorities would go so far to discredit it lends some evidence as to the list's authenticity.

64. The allusion here is to a case then very much in the news about a man killed reportedly after his vehicle was hit by a Molotov cocktail thrown by Shiʿa rioters in the southern village of Maʿameer. Seven were arrested and later handed life sentences in July 2010 under Bahrain's nebulous "anti-terrorism" statute of 2006. See Bahrain Centre for Human Rights, "Bahrain: Life Sentences against 7 Activists in the 'Maʿameer' Case after an Unjust Trial," July 11, 2010, available at www.bahrainrights.org/en/node/3175.

65. Quoted in Michael Slackman, "Sectarian Tension Takes Volatile Form in Bahrain," *New York Times*, March 28, 2009.

66. Quoted in Bill Law, "Riots Reinforce Bahrain Rulers' Fears," *Sunday Telegraph*, July 22, 2007.

67. Bahry, "The Socio-economic Foundations of the Shiite Opposition," p. 134.

68. See American Federation of Labor and Congress of Industrial Organizations (AFL-CIO), 2011, "Concerning the Failure of the Government of Bahrain to Comply with Its Commitments under Article 15.1 of the US-Bahrain Free Trade Agreement," April 21, 2011.

69. Khuri, *Tribe and State in Bahrain*, p. 48.

70. Ibid., p. 52.

71. Ibid., p. 48.

72. This period is examined in detail in Gengler, "Royal Factionalism, the Khawalid, and Securitization of 'the Shīʿa Problem' in Bahrain."

73. Khuri, *Tribe and State in Bahrain*, pp. 48, 94–95.

74. Cf. note 60.

75. Jill Crystal, "Patterns of State-Building in the Arabian Gulf," PhD diss., Harvard University, 1986; and Jill Crystal, *Oil and Politics in the Gulf: Rulers and Merchants in Kuwait and Qatar* (New York: Cambridge University Press, 1990).

76. That is, rejectors (*rawāfiḍ* sing. *rāfiḍī*) of the "rightful" successors of the Prophet in favor of his cousin and son-in-law 'Ali ibn Abi Talib; or, less specifically, rejectors of Sunni Islamic leadership and authority. The word is an offensive religious slur.

77. Muhammad Ghanim al-Rumaihi, *Bahrain: Social and Political Change since the First World War* (London: Bowker, 1978), p. 26.

78. Ibid., p. 25.

Chapter 3

1. Indeed, the two communities speak entirely different dialects of Arabic. See chapter 2, p. 42 and chapter 4, p. 90.

2. Though beyond the scope here, also worth noting in this regard is newspaper subscription boxes as indicators of political allegiance. An *Al-Wasat* (*The Center*) box, for example, affixed to many a Shi'a home, is a clear indicator of opposition support. On the other hand, *Al-Ayam* (*The Days*), *Al-Bilad* (*The Country*), and *Akhbar Al-Khaleej* (*The Gulf News*), are all safe pro-government choices, the first liberal-leaning but owned by a former minister of information turned advisor to the king; the latter two close to the prime minister. For a more hard-line statement one can opt for *Al-Watan* (*The Nation*), a mouthpiece of the royal court and often inflammatory.

3. The labyrinthine system of security checkpoints erected after 2011 has militated somewhat against this practice, as Shi'i drivers seek to avoid attracting unnecessary police attention.

4. And, as if to be even more emphatic in this point, the vowel in the word (*Āl*) that refers to the "family" of the Prophet is elongated for several seconds for each of three recitations, producing an affecting meter in which every syllable is deliberately uttered: "*allaaa / hum sal-lī / wa sal-lim / 'alaa muhammad / wa aaaaaaaaal muhammad.*"

5. Quoted in Habib Toumi, "Minister: Don't Use Religious Events to Fuel Sectarianism," *Gulf News*, February 15, 2007.

6. That is, the "Party of the Islamic Call." Founded in 1957 in Najaf, today it comprises one-half of the United Iraqi Alliance bloc led by former Iraqi prime minister Nuri al-Maliki. For more on the history of al-Da'wa in Iraq and the activities of its Bahrain wing, see Lawrence Louër, *Transnational Shia Politics* (New York: Columbia University Press), pp. 83–88ff.

7. All quoted in Habib Toumi, "Bahrain's Islamist MP Calls for Removal of Sectarian Banners," *Gulf News*, February 19, 2006.

8. Personal interview, Bahrain, May 2009. It was for precisely this sentiment, and his rejection of any compromise with the ruling family, that Sh. 'Abd al-Wahhab Husain was arrested and sentenced to life in prison in early 2011.

9. Fuad Khuri, *Tribe and State in Bahrain* (Chicago: University of Chicago Press), pp. 73–74.

10. Cf. chapter 2, note 40.

11. Khuri, *Table and State in Bahrain*, p. 76.

12. Louër, *Transnational Shia Politics*, p. 208.

13. See "ظلمتينا وكم كنتي ظلومة للبحرينين" ["'Oh How You Oppressed Us!,' for Bahrainis"], available at www.youtube.com/watch?v=-aS6eu1RfTc.

14. Haydar ("lion") is one of the nicknames of the Imam 'Ali and a common given name among Shi'i males.

15. This intra-Shi'a disagreement gives occasion to highlight another important aspect of Ashura, which if less central to the present discussion bears mentioning nonetheless. Just as the ritual reiterates the magnitude of the Sunni-Shi'i schism, so too does it serve to throw into stark relief the many factions of Shi'ism in Bahrain: Persian Shi'a vs. the Arab Baharna; al-Wifaq followers vs. supporters of the more radical opposition; and the majority adherents of Khomeini's *vilāyat-e faqīh* (rule by the "Guardian Jurist") doctrine vs. the so-called Shirazi faction. (See Louër's *Transnational Shia Politics* for the history of this latter Najaf-Karbala rivalry and its transportation to the Arab Gulf.) These overlapping divisions, finally, are formalized in the institution of the *ma'tam*, whose membership revolves primarily around such ethnic, political, and jurisprudential distinctions, and which compete against each other for the largest and most elaborate *'azza'* processions, which they organize. On this see Khuri, 1980, *Tribe and State in Bahrain*, chapter 8.

16. Khuri, *Tribe and State in Bahrain*, p. 77.

17. Indeed, it is no coincidence that Khomeini himself chose this exact date to voice his first attack on the Shah in June 1963 during the so-called Khordad uprising (Cf. Louër, *Transnational Shia Politics*, p. 187, note 32). The mass street protests of December 1978 that led to the downfall of the Iranian regime some two months later began on the twelfth of Muharram, spurred on by an oral communiqué issued by Khomeini on November 23 titled "Muharram: The Triumph of Blood over the Sword," which opened thusly (in *Islam and Revolution*, Hamid Algar, trans. [New York: Kegan Paul, 2002], p. 242):

> With the approach of Muharram, we are about to begin the month of epic heroism and self-sacrifice—the month in which blood triumphed over the sword, the month in which truth condemned falsehood for all eternity and branded the mark of disgrace upon the forehead of all oppressors and satanic governments; the month that has taught successive generations throughout history the path of victory over the bayonet.

18. For more about the incident, see Human Rights Watch, "Bahrain: Activist Jailed after Criticizing Prime Minister," September 28, 2004, available at www.hrw.org/english/docs/2004/09/29/bahrai9413.htm.

19. The video has since been uploaded to YouTube and features a quite heated argument in the comments section. See "الناشط عبد الهادي الخواجة: فالنسقط العصابة الحاكمة" ["Activist 'Abd al-Hadi al-Khawajah: 'Let's Take Down the Ruling Gang'"], available at www.youtube.com/watch?v=rC8ANWoUarU.

20. 'Abd al-Hadi al-Khawajah, "تضحيات الحسين تفضح 'العصابة الحاكمة' وتسقطها من الحكم" ["How the Sufferings of al-Husain Exposed 'the Ruling Gang' and Toppled It from Power"], unpublished address delivered in Manama near the al-Khawajah Mosque, January 7, 2009. All subsequent quotations are translated from an unpublished Arabic text of the speech compared to the public video.

21. It is ironic of course that people from "all streams and sects" should be called to action in the name of such a quintessentially Shi'i figure.

22. That is, the dynasty to which Yazid belongs, known collectively as the "Sufyanids" after his grandfather Abu Sufyan.

23. The notion of "group feeling" born of tribal co-sanguinity attributed to 14th-century Arab historian Ibn Khaldun. Cf. his *Al-Muqaddima*, 1, 234. For a modern application, see Ghassan Salamé, "'Strong' and 'Weak' States: A Qualified Return to the *Muqaddimah*," in Giacomo Luciani, ed., *The Arab State* (Berkeley: University of California Press, 1990), which analyzes the central role of *'aṣabiyya* in the unification of the Arabian Najd under Ibn Sa'ud.

24. A son of Yazid and one of his military commanders. According to the common Shi'i account, al-Hurr was charged with obstructing al-Husain's passage near Kufah but instead was convinced of his cause and defected to his side.

25. For additional emphasis the conclusion of the second line—"in dispute with a criminal"— was augmented by the interjection: "a criminal that is in the palace; the criminals that are [living] in the palaces!"

26. Cf. note 19.

27. Interview with 'Abd al-Jalil al-Singace, Bahrain, April 2009. The political spokesman of the al-Haqq Movement, al-Singace remains among Bahrain's most identifiable opposition figures. In August 2010 he was arrested upon his return from a British parliamentary session on human rights in Bahrain, accused of heading a "terrorist network." Eventually released in 2011, he was re-arrested only days later for his alleged involvement in organizing anti-government protests in February and March.

28. Personal interview, Bahrain, April 2009.

29. These pardons by the king set off what might be termed a mild controversy. Though glad to see the release of so many political detainees, most Shi'a cynically chalked up the gesture to Bahrain's fast-approaching Formula One Grand Prix race, whose foreign spectators were unlikely to be impressed at the sight of thousands of protesters along the main highways leading to the track. Sunnis, for their part, were largely critical of the leniency shown to these troublemakers, with some even sensing dissension within the Al Khalifa ranks. One Sunni member of parliament with whom I spoke claimed that the prime minister personally opposed the action, as did Saudi Arabia, where the premier visited the very day before the pardons. According to this account, the Saudi king made his displeasure known in a letter to King Hamad, and then by temporarily halting the passage of some 300 trailer trucks bound for Bahrain at the Saudi side of the causeway. Interview with 'Isa Abu al-Fath, Bahrain, April 2009. See Anon., "عفوٌ ملكيّ عن 178 محكوماً بقضايا أمنية وسياسية . . . والبحرين تبتهج" ["A Royal Pardon of 178 Convicted on Security- and Political-Related Cases . . . and Bahrain Rejoices"], *Al-Wasat*, April 18, 2009.

30. Louër, *Transnational Shi'a Politics*, p. 237.

31. Ibid., p. 290.

32. Ibid., pp. 292–293.

33. Ibid., p. 290.

34. Personal interview, Bahrain, May 2009.

35. Personal interview, Bahrain, April 2009.

36. Personal interviews, Bahrain, April and May 2009.

37. In the aftermath of the uprising some registered societies, such as Wa'ad and the Islamic Acton Society, were disbanded anyway for their support of mass protests. Post-uprising factions such as the village-based February 14th Youth Coalition are fundamentally decentralized.

38. Personal interview, Bahrain, May 2009.

39. Habib Toumi, "Religious Fatwas Used to Explain Poll Participation," *Gulf News*, November 21, 2006.

40. One might also cite al-Wifaq's seeming political miscalculations during negotiations in the early days of unrest. See Jane Kinninmont, "Bahrain: Beyond the Impasse: Programme Report" (London: Royal Institute of International Affairs), p. 5.

41. As discussed in chapter 1, notes 11 and 22, this argument originates in Skocpol's work on Iran. It is thereafter taken up in F. Gregory Gause, "Regional Influences on Experiments in Political Liberalization in the Arab World," in Rex Brynen, Bahgat Korany, and Paul Noble, eds. *Political Liberalization and Democratization in the Arab World*, Vol. 1, *Theoretical Perspectives* (Boulder, CO: Lynne Rienner, 1995); John Clark, "Petro-politics in Congo," *Journal of Democracy* 8.3 (1998): 62–76; and James D. Fearon, "Primary Commodity Exports and Civil War,"

Journal of Conflict Resolution 49.4 (2005): 483–507; and is tested along with many others by Ross. Finally, it receives extensive treatment by Fearon and Laitin, who appeal to repression to explain the incidence (and, in Bahrain, non-incidence) of civil war in countries heavily reliant upon primary commodity export rents. [James D. Fearon and David Laitin, "Bahrain," Stanford University, 2005, unpublished manuscript available at www.stanford.edu/group/ethnic /Random Narratives/BahrainRN1.1.pdf.]

Chapter 4

1. The case of Emirati-Iranian relations is complicated, however, by a longstanding dispute over three islands in the waters bordering the two countries claimed by the UAE but controlled by Iran.

2. Gwenn Okruhlik, "Rentier Wealth, Unruly Law, and the Rise of Opposition: The Political Economy of Oil States," *Comparative Politics* 31.3 (1999): 295, 297.

3. Indeed, one prominent Gulf scholar then at the London School of Economics, Kristian Coates Ulrichsen, was even famously denied entry to the United Arab Emirates, apparently at Bahrain's urging, because he was due to present a paper on the domestic and regional implications of the February 14th uprising. He discusses the episode, as well as the larger issue of academic free in the Gulf region, in Kristian Coates Ulrichsen, "Academic Freedom and UAE Funding," *Foreign Policy*, February 25, 2013.

4. A June 2009 article in the opposition-affiliated independent newspaper *Al-Wasat* reports that the BCSR was "liquidated" by virtue of Royal Decree No. 52 of 2009, its employees being "distributed among the ministries" and its building taken over by a "Center for Strategic and Energy Studies." See Anon., "حل مركز البحرين للدراسات والبحوث وتوزيع الموظفين على الوزارات" ["The Dissolution of the Bahrain Center for Studies and Research and Distribution of the Staff amongst the Ministries"], *Al-Wasat*, June 15, 2009.

5. It must be noted that the U.S. government as represented by its embassy seemed also to share this anxiety, not only about the political implications of the execution of such a project, but about those too of its likely findings. At one point after the fallout with the BCSR, in fact, the question was raised whether I should be altogether proscribed from administering the survey. While it was decided at last that U.S. embassies are not in the business of blessing or striking down the academic pursuits of citizens, and though many there showed much-appreciated support for the project, when I finally left Bahrain I was asked with keen interest when I anticipated the results would be made public.

6. James D. Fearon and David Laitin, "Bahrain," Stanford University, 2005, p. 8, unpublished manuscript available at www.stanford.edu/group/ethnic/Random Narratives /BahrainRN1.1.pdf.

7. Clive Holes, "Dialect and National Identity: The Cultural Politics of Self-Representation in Bahraini *Musalsalāt*," in Paul Dresch and James Piscatori, eds., *Monarchies and Nations: Globalization and Identity in the Arab States of the Gulf* (London: I. B. Tauris, 2005), p. 60.

8. If one wondered before at the decision to oversee the survey myself after the problems with the BCSR, rather than outsource the job to another local contractor, perhaps the foregoing may serve as an explanation. As there is no private Bahraini alternative to the BCSR, the only other options would have been two foreign-based market research companies: the Pan Arab Research Center and the Market Research Organization, the latter being employed, I was told, by the U.S. embassy for its internal Bahrain-related polling. But after communicating with the

heads of both organizations, which in any case were quite backlogged with consumer survey work, I concluded that each would be ill-suited for my needs: a majority of their field interviewers were foreigners; they could not be expected to be attentive to the political sensitivities and idiosyncrasies of the various regions of Bahrain, above all in the Shiʿa villages; and finally it seemed advantageous to have a personal relationship with those conducting the interviews, which would not only afford me more control over the interviewing process, but also allow me a window into the more intimate details of the interviews by hearing individual anecdotes and experiences in the field, such as have been related already. Finally, this hands-on approach meant that I could be certain of the quality of the sample utilized for the survey, about which more will be said shortly.

9. The only political current one might consider absent, then, is the Shiʿa Islamic Action Society headed by (the now-imprisoned) Sh. Muhammad ʿAli al-Mahfudh. The direct modern descendent of the Islamic Front for the Liberation of Bahrain best known for its alleged failed coup attempt of 1981, the group adheres to the Shirazi school of theology and, as such, is fundamentally at odds with al-Wifaq, in principle competing with it for support. After boycotting the 2002 elections, however, its candidates failed to win any seats in 2006 and thus became a relative political non-player even prior to its 2012 dissolution. Moreover, the al-Khawajahs being among the most prominent of the Shirazi families of Bahrain, my interview with ʿAbd al-Hadi perhaps serves to compensate for this omission.

10. This was less so of the business elite of Muharraq, the historical seat of Bahrain's government until its move to Riffa on the main island in 1923. Bahrain's second-largest city and only city-governorate other than Manama, Muharraq is widely acknowledged as home to "the most politically aware" citizens, that is, Sunni citizens, in the country, as said by one of its leading merchants. In practice this means that the natives of this traditionally Sunni stronghold are relatively less reserved in voicing criticism of the government and of the ruling family than their more tribally aligned counterparts in, for instance, Riffa. Such is particularly the case on the issues of naturalization and economic management and policy, as both impact their business interests.

11. This interviewer effect is examined in some detail in the concluding chapter.

12. Beyond this, the fact that the sample was destined originally for use in one of the BCSR's own, state-sponsored surveys before being passed to me makes it very unlikely that it would have been doctored or truncated.

13. Block numbers beginning with 100 correspond to the area of al-Hidd; the 200s to Muharraq; 300s to Manama and the island of Nabih Salih; 400s to Jidd Hafs and several Shiʿa villages; 500s to the northern region dominated by Shiʿa villages; 600s to the Shiʿa stronghold of Sitra; 700s to a Shiʿa central region; 800s to Isa Town; 900s to the Sunni tribal stronghold of Riffa and the sparsely populated, militarized southern two-thirds of the island; 1000s to a western region inhabited only in a few coastal villages; the 1100s to the Hawar Islands; and the 1200s to the demographically mixed Hamad Town, the country's newest urban development and home to many naturalized Sunnis.

14. Kingdom of Bahrain, "Population of the Kingdom of Bahrain by Governorate, Nationality & Sex—2010," Central Informatics Organization.

15. Anon., "المعاودة: 318668 مجموع الكتلة الانتخابية للعام 2010" ["al-Maʿawdah: 318,668 Is the Total Electoral Bloc for 2010"], *Al-Wasat*, September 9, 2010. Though *Al-Wasat* elsewhere contradicts this official government count, its own estimate of 19.0 percent (60,906 of 321,000) is not so different. See Anon., "سادسة الجنوبية الأقل بـ770 ناخباً وأولى الشمالية الأكبر بـ16223" ["The Southern Sixth [District] Is the Smallest with 770 Electors and the Northern First Is the Biggest with 16,223"], *Al-Wasat*, August 25, 2010.

16. The large representation of the urban Manama and Muharraq neighborhoods, which together account for nearly half of the uncompleted interviews, perhaps implies a slight disproportionate omission of Sunnis. For the interested reader, the unrepresented and under-represented regions correspond to the following areas and block numbers: 209 (Muharraq Town), 213–214 (Muharraq Suq), 216 (South Muharraq), 301 (Manama Suq), 306 (Ra'as Ruman), 314 (al-Nu'aim), 318 (al-Hurra), 321 (Gudaibiyya Suq), 408 (Sanabis), 419 (Jidd Hafs), 430 and 434 (Karbabad), 433 (Jablat Hibshi), 436 (Seef), 526 (Barbar), 542 (al-Diraz), 551 (al-Gharbiyya), 555 (al-Budayyi'), 561 (al-Janabiyya), 623–624 (East and West 'Akar), 633 (Ma'ameer), 644 (Nuwaidrat), 721 (Jidd 'Ali), 1010 and 1014 (al-Hamalah), and 1101 and 1103 (Hawar Islands).

17. The site previously existed at www.bahrainexplorer.com/bex. The BCSR's parent organization, the Central Informatics Organization, now seems to offer a similar interactive GIS map at www.bahrainlocator.gov.bh.

18. Qubain tells that the census was taken "primarily for food control purposes." ("Social Classes and Tensions in Bahrain," *Middle East Journal* 9.3 [1955]: 269.)

19. The use of governorate-level sampling probability weights increases the estimated Shi'i proportion only slightly to 58.2 percent, with a corresponding 95 percent confidence interval of between 54.1 percent and 62.3 percent.

20. In September 2014, the Central Governorate was dissolved as part of extensive electoral redistricting, its constituent districts disbursed among the remaining four regions.

21. "Bahrain—Population Density," *Atlas of the Middle East* (Washington, DC: U.S. Central Intelligence Agency, 1993), available at www.lib.utexas.edu/maps/atlas_middle_east/bahrain_pop.jpg.

22. "Population Density Map of Bahrain," *Best Country Reports* (Petaluma, California: World Trade Press, 2007), available at www.bestcountryreports.com/Population_Map_Bahrain.html.

23. These results come from the most recent 2010 parliamentary elections, in which al-Wifaq gained one seat (in a central Manama district) to increase its total to 18. Using the 2006 election results would not have substantially altered the contours of Map 4.3, therefore. For a summary of the 2010 results, see Anon., "Bahrain's First Round Parliamentary Election Results," *Gulf News*, October 24, 2010.

24. It is a clear testament to the shrewdness of the Bahraini public relations department—and the Western public relations firms employed by the government—that the district won in 2006 and 2010 (both times unopposed) by the celebrated "first female MP in the Gulf"—the Southern Sixth—consists of the unpopulated Hawar Islands along with mainland blocks 998, 997, 973, 971, 967, and 965, of which only the last is home to any considerable population, and which in total accounted for just 770 registered voters in the 2010 election. See Anon., "سادسة الجنوبية الأقل بـ770 ناخباً وأولى الشمالية الأكبر بـ16223" ["The Southern Sixth [District] Is the Smallest with 770 Electors and the Northern First Is the Biggest with 16,223"], *Al-Wasat*, August 25, 2010. Presumably to counter such criticism, the district was expanded somewhat to include new residential neighborhoods in the electoral redistricting of September 2014. See note 20.

For this—as, for instance, the country's 2008 appointment to the United States of the first female, Jewish ambassador of an Arab nation from among its 36-strong Jewish community—one is impressed at the scale of Bahrain's political maneuvering. As then matter-of-factly explained by the head of the Bahrain Human Rights Society, "We always believe here that control of America is governed by the Zionist lobby. The media and the money are all in the hands of the Jews. We believe if we have a Jewish ambassador and Jews in the Shura Council, this is a positive indicator for the country." See Michael Slackman, "In a Landscape of Tension, Bahrain Embraces Its Jews. All 36 of Them," *New York Times*, April 5, 2009.

25. By governorate, these voter averages are: 5,146 Shi'i to 3,294 Sunni in the Capital; 7,182 to 7,675 in Muharraq; 13,107 to 7,631 in the Northern; 12,699 to 9,470 in the Central; and the average Southern Governorate district, all six of which were easily carried by Sunni MPs, contained a mere 2,913 electors in 2010.

26. Though many Shi'a would claim that they are effectively barred from owning property in Sunni-dominated areas such as Riffa and parts of Muharraq.

Chapter 5

1. Hamad bin 'Isa Al Khalifa, "Stability Is a Prerequisite of Progress," *Washington Times*, April 19, 2011.

2. Respondents were asked exactly this, i.e., "What is your sector of work: public or private?"

3. James J. Heckman, "The Common Structure of Statistical Models of Truncation, Sample Selection, and Limited Dependent Variables and a Simple Estimator for Such Models," *Annals of Economic and Social Measurement* 5.4 (1976): 475–492.

4. More generally, the Heckman model offers a correction for sample selection bias by modeling the selection process using data on those *not selected*, treating the problem as if it were one of an omitted variable. The selection model is estimated by probit, producing estimated inverse Mills ratio values for each selected case. Using these estimated values, the behavioral model is then estimated by a generalized least-squares regression of Y on the Xs and estimated inverse Mills ratios. This procedure gives unbiased estimates for B and an estimated ρ, the correlation between the error terms in the behavioral and selection models.

The models here are estimated using Stata's heckman and heckprob implementations, as appropriate. Maximum likelihood estimates are used rather than Heckman's two-step estimates, as the latter is not compatible with weighted survey data.

5. The original seven-point scale was recoded so that the small number of observations (<20) in the extreme categories would not unduly influence estimates.

6. As indicated in the model specification, martial status is interacted with the *female* indicator to capture the fact that married females are more likely to be (unemployed) stay-at-home parents than married males.

7. The regressor age^2 ($age \times age$) is also included to model properly the quadratic rather than linear relationship between age and employment.

8. Including an additional *female* × *sect* term to control for a possible Sunni-Shi'i difference in workforce participation among females does not alter the results of the selection or behavioral model.

9. Predicted probabilities of employment calculated at overall (not group-specific) means of gender, age, and martial status controls.

10. If one excludes working police and military personnel from this analysis—all twelve of whom, as described shortly, are Sunnis and disproportionately of below average education— the overall Sunni-Shi'i gap in public sector employment remains, if at a reduced statistical and substantive significance. In this case, the estimated conditional probability of government employment among Sunnis falls to 56 percent, with a Sunni now estimated to be around 31 percent more likely (in relative terms) to be employed with the government than a Shi'i of the same age, marital status, gender, and education level. This overall Sunni-Shi'i difference is significant only at the $p=0.119$ level.

However, even with police and military excluded, the sectarian discrepancy remains highly significant ($p=0.003$) among those of the lowest education category, representing almost 15 percent of all working-age Bahrainis. An estimated 10 percent gap in probability of public-sector employment continues to separate Sunnis and Shi'is of the terminal secondary education category, but this difference is not statistically significant when police and military personnel are excluded.

11. The exact categories were: (1) "employer/manager of establishment with 10 or more employees"; (2) "employer/ manager of establishment with less than 10 employees"; (3) "professional worker (lawyer, accountant, teacher, etc.)"; (4) "supervisory office worker"; (5) "non-manual, non-supervisory office worker"; (6) "foreman, supervisor"; (7) "skilled manual worker"; (8) "semi-skilled manual worker"; (9) "unskilled manual worker"; (10) "farmer, owns farm"; (11) "agricultural worker"; (12) "member of the armed forces, police"; and (13) "housewife." See Figure 5.3.

12. One female Shi'i respondent did report that her spouse worked for the police or armed forces, but since he was not interviewed one cannot be certain of his sectarian affiliation. The same applies to the analogous discussion of Sunni spouses below.

13. Moreover, two older Sunnis report being members of the armed forces or police but indicate that they are not currently working, so these two observations are here excluded.

14. In practice, the exclusion of these categories, whether together or individually, does not substantively alter the results. It offers in fact a more conservative estimate of the effect of sectarian membership on job level.

15. Among Shi'a respondents under 60 (retirement age) who do not report being a housewife, approximately 9 percent hold a primary diploma or less, 28 percent a secondary diploma, and 19 percent have some post-secondary education below the level of a bachelor's degree.

Of course, one might argue that the underrepresentation of lower-educated Shi'a relative to Sunnis in the public sector is at least in part a result of job-seekers' behavior beyond that of employers. One might hypothesize, say, that lower-educated Shi'a are more likely than Sunnis to begin their own business or to join a family business. It might be, in other words, that the observed Sunni-Shi'i discrepancy in sector of employment reflects actually a relative *preference* among the latter for work in private industry. Yet, such an argument must offer an *a priori* explanation not only for the source of this preference, but also for why it exists only among lower-educated individuals. Furthermore, the simultaneous sectarian discrepancy in job status among less-educated individuals must also find a consistent explanation.

16. See the second paragraph of note 15.

17. One might wonder why we would resort in gauging the strength of a Bahraini's confessional identity to a general measure of religiosity as opposed to, say, by asking a respondent directly about communal attachment. Yet, because confessional membership was, to avoid further raising the suspicious and apprehensions of respondents, inferred rather than inquired about directly, this was not an option.

18. By this measure, 268 of 401 (or 69 percent) of total respondents are coded as "religious," and 133 as "not religious." Compare this to the 221 of 389 (or 57 percent) of individuals who self-identified as "religious" when asked directly. (If one includes those who replied "moderate" this increases to 71 percent). Yet the correlation between the two measures is a relatively low 0.263 (or 0.295 if the "moderates" are excluded), meaning that on the whole the sort of person who self-identifies as "religious" when asked directly is not the same sort of person who exhibits concern for the religious in the case of what is in the Arab world among life's most significant practical matters: marriage.

19. This is all the better, as around one-third of respondents declined to give an estimate of total household income. Only seven individuals—five Shi'is and two Sunnis—reported a "very

bad" household economy, so to better facilitate a categorical-based analysis these responses are combined to form a single "bad" category.

20. As far I am aware, this represents the first systematic study of sectarian interviewer effects in the Arab world.

21. Justin Gengler, "Understanding Sectarianism in the Persian Gulf," in Larry Potter, ed., *Sectarian Politics in the Persian Gulf* (London/New York: Hurst/Oxford University Press, 2014).

22. As measured by the R^2 statistic for the corresponding bivariate OLS regression. The bivariate correlation between the two variables is an impressive 0.63.

23. Justin Gengler, "Bahrain's Sunni Awakening," *Middle East Research and Information Project*, 2012.

24. While the latter effect is not as statistically robust as one would prefer, being significant only around the $p=0.10$ level, still its consistency in magnitude and direction across a variety of models of political opinion and behavior gives one confidence that its relative lack of statistical significance compared to among Shi'a owes principally to the smaller number of Sunnis in the survey sample. A diagnostic test offers some evidence that this is the case: when one drops the interaction terms involving the demographic controls, replacing the usual Sunni- and Shi'i-specific estimates with a single estimate and so gaining several degrees of freedom, the statistical significance of the effect among Sunnis increases by some 50 percent.

25. For instance, all but 15 of 1,060 citizens surveyed for the 2010 Qatar World Values Survey (WVS) said they were "very proud" to be Qatari. In the 2003 WVS survey of Saudi Arabia, this proportion was more in line at least with Bahraini Sunnis, at 73 percent of respondents.

26. The political nature of the former two activities is made explicit in the Arabic wording of the questions. The former asks whether a respondent has participated in a demonstration (*muẓāhara*) or march (*masīra*) in the previous three years. The latter asks if a respondent has signed a petition or attended a meeting "in order to discuss an issue."

Due to a relatively high rate of non-response for these questions owing to their political sensitivity, the group-specific estimates for the demographic controls here are replaced by single estimates in order to gain additional statistical leverage. That is, the interaction terms involving *age*, *education*, and *female* are dropped. Cf. Figure 3 in the appendix.

27. As indicated by the error bars representing the 95 percent confidence intervals for the respective estimates, the statistical significance of these effects among Sunnis is relatively less robust than among Shi'a. As elsewhere, this owes in large part to the relatively smaller number of Sunni respondents. Cf. notes 24 and 26.

28. Strictly speaking, only the difference in estimated participation between the first and third categories of economy is statistically significant at around the $p=0.05$ level.

29. Personal interview, Bahrain, May 17, 2009.

30. Personal interview, Bahrain, April 19, 2009.

31. Fuad Khuri, *Tribe and State in Bahrain* (Chicago: University of Chicago Press, 1980), p. 225.

32. Quoted in Anon., "Bahrain's Pre-election Jitters," *The Economist*, October 14, 2010.

33. Quoted in Gengler, "Bahrain's Sunni Awakening," *Middle East Research and Information Project*, 2012.

34. Al-Wifaq National Islamic Society, "Opposition Societies Condemn the Arrest of Mohamed Al Zayani and Demand His Immediate Release," July 21, 2010, available online at http://alwefaq.net/cms/2012/07/21/6695.

35. Donna Abu-Nasr, "Business Owned by Bahraini Lawmaker Riddled with Bullets," *Bloomberg*, April 29, 2012.

36. This period is treated in much greater detail in Khuri, *Tribe and State in Bahrain*, pp. 24ff.; and Justin Gengler, "Royal Factionalism, the Khawalid, and the Securitization of 'the Shī'a Problem' in Bahrain," *Journal of Arabian Studies* 3.1 (2013): 53–79.

37. Daniel Brumberg, "Transforming the Arab World's Protection-Racket Politics," *Journal of Democracy* 24.3 (2013): 88–103.

Chapter 6

1. Anon., "Premier's tribute . . . ," *Gulf Daily News*, April 22, 2011.

2. See, e.g., Michael Slackman, "Bahrain's Sunnis Defend Monarchy," *New York Times*, February 17, 2011.

3. Fuad Khuri, *Tribe and State in Bahrain* (Chicago: University of Chicago Press), pp. 232–233.

4. In Thanassis Cambanis, "Crackdown in Bahrain Hints of End to Reforms," *New York Times*, August 26, 2010.

5. Recall, moreover, that included in the models is an additional variable, *refuse*, meant to control for exactly this general apprehension. Cf. chapter 5, p. 122.

6. A few methodological notes are in order. A first is that the question about the 2006 elections was asked of all citizens independent of whether they actually participated in it. And, in fact, 77 percent of Shi'a and 58 percent of Sunnis who report not having voted still give an evaluation of its legitimacy. Besides being interesting per se, this fact is important as it shows that the distribution of observations is not truncated by a selection effect such that one observes opinions about the election only by those who voted.

A second note relates to the sectarian interviewer effect generally. In particular, one might suggest that the effect owes perhaps not to the interviewer, but rather to the nature of the mixed-sect neighborhoods where these interviews took place. That is, one might argue that, since interviewers were sent to locations with a view to their sectarian composition, individuals most likely to be interviewed by a member of the opposite confessional group naturally will tend to reside in these more diverse neighborhoods, and thus may have more moderate political views than those living in more isolated, exclusive areas. Such a result would amount, then, to a selection effect rather than an interviewer effect.

In reality, however, less than half of mixed-sect interviews took place in mixed-sect neighborhoods. Moreover, apart from the historically less politicized city of Isa Town, many or most of Bahrain's more mixed areas, such as, for instance, Hamad Town, and the border between the Sunni enclave of al-Budayyi' village and the many neighboring Shi'a villages, are sites of extreme contestation and at times communal violence.

Finally, for illustrative purposes the model estimation here is done by OLS regression despite the more categorical nature of the dependent variable.

7. In Thomas Fuller, "Bahrain's Promised Spending Fails to Quell Dissent," *New York Times*, March 6, 2011. The quote comes from Wa'ad leader Ebrahim Sharif.

8. Mohammed al-A'ali, "Penalties plan for not voting," *Gulf Daily News*, November 21, 2014. Quoted in Justin Gengler, "Electoral rules (and threats) cure Bahrain's sectarian parliament," *Washington Post*, December 1, 2014.

9. According to the authoritative database compiled by the Stockholm International Peace Research Institute, between 2000 and 2009 the top 11 military spenders as a proportion of GDP include five of the six GCC states: Oman (#1), Saudi Arabia (#2), the UAE (#4), Kuwait (#6),

and Bahrain (#11). Data for Qatar are not reported for the years 2000, 2001, and 2009, but based on the incomplete data it would rank at #30. The data are available at www.sipri.org/databases /milex.

10. See note 9.

11. Shafeeq Ghabra, "Kuwait and the Economics of Socio-economic Change," in Barry M. Rubin (ed.), *Crises and Quandaries in the Contemporary Persian Gulf* (New York: Frank Cass, 2002), p. 112.

12. Mariam Al Hakeem, "Thousands in Saudi Arabia after Losing Qatari Citizenship," *Gulf News*, April 3, 2005; and Mariam Al Hakeem, "Citizenship Restored to 5,266 Qataris," *Gulf News*, February 3, 2006.

13. Rania El Gamal, "UAE Detains 6 Islamists Stripped of Citizenship: Lawyer," *Reuters*, April 9, 2012.

14. Saad Abedine and Mustafa Al-Arab, "Bahrain Strips Shiite Activists of Citizenship amid Unrest," *CNN*, November 8, 2012.

15. Decree 20 of July 31, 2013. Text (in Arabic) available via the Bahrain News Agency at www .bna.bh/portal/news/573609.

16. Kareem Fahim, "Kuwait, Fighting Dissent Within, Revokes Citizenship," *New York Times*, September 30, 2014.

17. Personal interview, former al-Wifaq MPs, Doha, May and October 2013.

18. According to Secretary General Sh. ʿAli Salman, after studying the changes the bloc concluded that it could win, as in 2010, no more than 18 of the 40 districts. See Malik Abdullah, "سلمان: الدوائر الجديدة ثبّتت أغلبية فوز المقربين للسلطة بـ22 مقعداً" ["Salman: New [Electoral] Districts Guarantee Majority for [Candidates] Close to the Authorities, with 22 Seats"], *Al-Wasat*, September 27, 2014.

19. For instance, the Crown Prince–linked advocacy group Citizens for Bahrain notes in its comprehensive overview of the electoral district changes, "[I]t is unclear whether the societies belonging to the Al-Fateh Coalition will succeed in forming a political bloc. The change in constituency boundaries seems to have complicated this process." Anon., "Implications of redrawn constituency borders in Bahrain," *Citizens for Bahrain*, October 2, 2014, available at www.citizensforbahrain.com/index.php/featured-articles/entry/implications-of-redrawn-co nstituency-borders-in-bahrain.

20. Alaʾa Shehabi, "Why is Bahrain Outsourcing Extremism?" *Foreign Policy*, October 29, 2014.

21. Ibid.

Appendix

1. Owing to the coding of the *sect* dummy indicator, which takes on a value of 1 for Sunnis, the coefficients and significance levels for interaction terms involving *sect* (e.g., *econ2 × sect*) give the effect of that term *among Sunni respondents*; whereas the estimates for the respective standalone variable (e.g., *econ2*) apply to Shiʿi respondents. For a practical guide to interpreting multiplicative interaction models, see, Thomas Brambor, William R. Clark, and Matt Golder, "Understanding Interaction Models: Improving Empirical Analyses," *Political Analysis* 14.1 (2006): 63–82. Note that all regressions make use of Stata's factor variable operators for factor (dichotomous and categorical) variables, continuous variables, and interactions.

Bibliography

A'ali, Muhammad, al-. 2008. "Session Disrupted over 'Bandargate.'" *Gulf Daily News*, March 12.

Abdullah, Malik. 2014. "سلمان: الدوائر الجديدة ثبّتت أغلبية فوز المقربين للسلطة بـ 22 مقعداً" ["Salman: New [Electoral] Districts Guarantee Majority for [Candidates] Close to the Authorities, with 22 Seats"]. *Al-Wasat*. September 27.

Abu al-Fath, 'Isa Ahmad. 2009. Personal interview. Bahrain. April 19.

Abu-Nasr, Donna. 2012. "Business Owned by Bahraini Lawmaker Riddled with Bullets." *Bloomberg*, April 29.

Ahmad, 'Ali. 2009. Personal interview. Bahrain. May 14.

Alhasan, Hasan Tariq. 2011. "The Role of Iran in the Failed Coup of 1981: The IFLB in Bahrain." *Middle East Journal* 65.4: 603–617.

Al Khalifa, Hamad bin 'Isa. 2011. "Stability Is a Prerequisite of Progress." *Washington Times*, April 19.

Al Khalifa, Mai bint Muhammad. 1999. *From the Surroundings of Kufa to Bahrain: The Carmathian, from an Idea to a State*. Beirut: Arab Institute for Studies and Publishing.

———. 2000. *Charles Belgrave: Biography and Diary, 1926–1957*. Beirut: Arab Institute for Studies and Publishing.

Al-Wifaq National Islamic Society. 2010. "Opposition Societies Condemn the Arrest of Mohamed Al Zayani and Demand His Immediate Release." July 21. Available online at http://alwefaq.net/cms/2012/07/21/6695.

American Federation of Labor and Congress of Industrial Organizations (AFL-CIO). 2011. "Concerning the Failure of the Government of Bahrain to Comply with Its Commitments under Article 15.1 of the US-Bahrain Free Trade Agreement." April 21.

Anderson, Benedict. 1983. *Imagined Communities: Reflections on the Origin and Spread of Nationalism*. London: Verso.

Anon. 2006. "BANDARGATE!" *Gulf Daily News*, September 24.

———. 2006. "BANDARGATE: The Unanswered Questions." *Gulf Daily News*, September 27.

———. 2008. "Kuwait MPs Expelled for Mourning Mughniyah." *Al-Arabiya*, February 20.

———. 2009. "عفوٌ ملكيّ عن 178 محكوماً بقضايا أمنية وسياسية . . . والبحرين تبتهج" ["A Royal Pardon of 178 Convicted on Security- and Political-Related Cases . . . and Bahrain Rejoices"]. *Al-Wasat*, April 18.

———. 2009. "شيعة السعودية يطالبون بحقوق أفضل في الجنوب: واليمن يتهم إيران بمحاولة إقامة دويلة شيعية." ["The Shi'a of Saudi Arabia Call for Better Rights in the South; and Yemen Accuses Iran of Trying to Create a Shi'a Mini-state"]. *Ma'rib Press*, May 27.

———. 2009. "حل مركز البحرين للدراسات والبحوث وتوزيع الموظفين على الوزارات" ["The Dissolution of the Bahrain Center for Studies and Research and Distribution of the Staff amongst the Ministries." *Al-Wasat*, June 15.

———. 2010. "المعاودة: 318668 مجموع الكتلة الانتخابية للعام 2010" ["al-Maʿawdah: 318,668 Is the Total Electoral Bloc for 2010"]. *Al-Wasat*, September 9.

———. 2010. "سادسة الجنوبية الأقل بـ770 ناخباً وأولى الشمالية الأكبر بـ16223" ["The Southern Sixth [District] Is the Smallest with 770 Electors and the Northern First Is the Biggest with 16,223"]. *Al-Wasat*, August 25.

———. 2010. "Bahrain's First Round Parliamentary Election Results." *Gulf News*, October 24.

———. 2010. "Bahrain's Pre-election Jitters." *The Economist,* October 14.

———. 2013. "GCC's Oil and Gas Annual Earnings Hit \$US737.5 Billion." *Oil Review Middle East*, March 19.

———. 2014. "Implications of redrawn constituency borders in Bahrain." Citizens for Bahrain. October 2, 2014. Available at www.citizensforbahrain.com/index.php /featured-articles/entry/implications-of-redrawn-constituency-borders-in -bahrain.

Arab Petroleum Research Center. 2004. *The Arab Oil and Gas Directory*. Paris: Arab Petroleum Research Center.

Asoomi Mohammad, al-. 2013. "Oman and Bahrain Have Lot to Gain from GCC Plan." *Gulf News*, June 19.

Atiqi, Suliman, al-. 2013. "Laboring against Themselves." *Sada*. Carnegie Endowment for International Peace. February 26.

Ayubi, Nazih. 1990. "Arab Bureaucracies: Expanding Size, Changing Roles." In Giacomo Luciani, ed., *The Arab State*. London: Routledge.

Bahrain Centre for Human Rights. 2002. "Documentary Film Script: The Political Naturalization in Bahrain." Available at www.bahrainrights.org/node/269.

———. 2006. "A Petition from a Hundred Prominent Figures and Activists to the King of Bahrain." October 13. Available at: www.bahrainrights.org/node/610.

———. 2006. "The Al Bander Report: What It Says and What It Means." Available at www.bahrainrights.org/node/528.

———. 2009. "Arbitrary Detention of a Citizen for Disseminating Information on the National Security Apparatus." June 21. Available at www.bahrainrights.org/en /node/2914.

———. 2010. "Bahrain: Life Sentences against 7 Activists in the 'Maʿameer' Case after an Unjust Trial." July 11. Available at www.bahrainrights.org/en/node/3175.

———. 2010. "Banning One of the Most Significant Historic Books in the History of Bahrain." May 25. Available at www.bahrainrights.org/en/node/3105.

Bahrain Independent Commission of Inquiry. 2011. *Report of the Bahrain Independent Commission of Inquiry*. November 23. Revised December 10. Available at www .bici.org.bh/BICIreportEN.pdf.

Bahrain-Saudi Arabia Boundary Agreement, February 22, 1958.

Bahry, Louay. 1997. "The Opposition in Bahrain: A Bellwether for the Gulf?" *Middle East Policy* 5.2: 42–57.

———. 2000. "The Socio-economic Foundations of the Shiite Opposition in Bahrain." *Mediterranean Quarterly* 11.3: 129–143.

Bandar, Salah, al-. 2006. "البحرين: الخيار الديموقرطي وآليات الإقصاء" ["Bahrain: The Choice of Democracy and the Machinery of Exclusion"]. Unpublished report prepared by the Gulf Centre for Democratic Development. Available (in Arabic) at www .bahrainrights.org/files/albandar.pdf.

Beblawi, Hazem. 1987. "The Rentier State in the Arab World." In Hazem Beblawi and Giacomo Luciani, eds., *The Rentier State: Nation, State and Integration in the Arab World*. Vol. 2. London: Croon Helm.

———. 1990. "The Rentier State in the Arab World." In Hazem Beblawi and Giacomo Luciani, eds., *The Arab State*. London: Routledge.

Belgrave, Charles. 1960. *The Pirate Coast*. Beirut: Librairie du Liban.

———. 1972. *Personal Column*. Beirut: Librairie du Liban.

Birnbaum, Ben. 2011. "Pro-government Cleric to Start Own Party in Bahrain." *Washington Times*, August 9.

Brambor, Thomas, William Roberts Clark, and Matt Golder. 2006. "Understanding Interaction Models: Improving Empirical Analyses." *Political Analysis* 14.1: 63–82.

Brumberg, Daniel. 2013. "Transforming the Arab World's Protection-Racket Politics." *Journal of Democracy* 24.3: 88–103.

Bueno de Mesquita, Bruce, James D. Morrow, Randolph M. Siverson, and Alastair Smith. 2003. *The Logic of Political Survival*. Cambridge, Mass.: MIT Press.

Burke, Edward. 2008. "Bahrain: Reaching a Threshold." Working paper presented at El Fundación para las Relaciones Internacionales y el Diálogo Exterior (FRIDE). Madrid. June 5. Available at www.fride.org/publication/452/bahrain-reaching-a -threshold.html.

Bushway, Shawn, Brian D. Johnson, and Lee Ann Slocum. 2007. "Is the Magic Still There? The Use of the Heckman Two-Step Correction for Selection Bias in Criminology." *Journal of Quantitative Criminology* 23.2: 151–178.

Chandra, Kanchan. 2004. *Why Ethnic Parties Succeed*. Cambridge, U.K.: Cambridge University Press.

Clark, John. 1998. "Petro-politics in Congo." *Journal of Democracy* 8.3: 62–76.

Corstange, Daniel M. 2008. "Institutions and Ethnic Politics in Lebanon and Yemen." PhD diss., University of Michigan.

Crystal, Jill. 1986. "Patterns of State-Building in the Arabian Gulf: Kuwait and Qatar." PhD diss., Harvard University.

———. 1990. *Oil and Politics in the Gulf: Rulers and Merchants in Kuwait and Qatar*. New York: Cambridge University Press.

Darwish, Adeed. 1999. "Rebellion in Bahrain." *Middle East Review of International Affairs* 3.1: 84–87.

Davidson, Christopher M. 2012. "The United Arab Emirates: Frontiers of the Arab Spring." *Open Democracy*, September 8.

Dawsari, Salman al-. 2009. "البحرين: نائب رئيس اللجنة التشريعية في البرلمان يتهم الوفاق بـ'تحركات' مشبوهة 'مع الحوثيين في اليمن" ["Bahrain: Vice-Chairman of the Legislative Committee in Parliament Accuses the al-Wifaq Opposition of 'Suspicious Dealings' with the Huthis of Yemen"]. *Al-Sharq Al-Awsat*, August 24.

Diwan, Kristin Smith. 2011. "Kuwait's Constitutional Showdown." *Foreign Policy*, November 17.

———. 2012. "Kuwait's Balancing Act." *Foreign Policy*, October 23.

Fahim, Kareem. "Kuwait, Fighting Dissent Within, Revokes Citizenship." *New York Times*. September 30, 2014.

Fakhro, Munira. 1997. "The Uprising in Bahrain: An Assessment." In Gary Sick and Lawrence Potter, eds., *The Persian Gulf at the Millennium: Essays on Politics, Economy, Security, and Religion*. New York: St. Martin's Press.

Fearon, James D. 2005. "Primary Commodity Exports and Civil War." *Journal of Conflict Resolution* 49.4: 483–507.

Fearon, James D., and David Laitin. 2005. "Bahrain." Unpublished manuscript available at www.stanford.edu/group/ethnic/Random Narratives/BahrainRN1.1.pdf.

Foley, Sean. 2010. *The Arab Gulf States: Beyond Oil and Islam*. Boulder, Colo.: Lynne Rienner.

Fuller, Thomas. 2011. "Bahrain's Promised Spending Fails to Quell Dissent." *New York Times*, March 6.

Gause, F. Gregory, III. 1995. "Regional Influences on Experiments in Political Liberalization in the Arab World." In Rex Brynen, Bahgat Korany, and Paul Noble, eds., *Political Liberalization and Democratization in the Arab World*. Vol. 1, *Theoretical Perspectives*. Boulder, Colo.: Lynne Rienner.

Gengler, Justin. 2011. "How Radical Are Bahrain's Shia?" *Foreign Affairs,* May 15.

———. 2012. "Bahrain's Sunni Awakening." *Middle East Research and Information Project*. January 17.

———. 2013. "Bahrain: A Special Case." In Fatima Ayub, ed., *What Does the Gulf Think about the Arab Awakening?* London: European Council on Foreign Relations.

———. 2013. "Royal Factionalism, the Khawalid, and the Securitization of the 'Shī'a Problem' in Bahrain." *Journal of Arabian Studies* 3.1: 53–79.

———. 2014. "Understanding Sectarianism in the Persian Gulf." In Larry Potter, ed., *Sectarian Politics in the Persian Gulf*. London/New York: Hurst/Oxford University Press.

Gray, Matthew. 2011. "A Theory of 'Late Rentierism' in the Arab States of the Gulf." *Occasional Paper No. 7*. Doha: Center for International and Regional Studies.

Groh, Mathew, and Casey Rothschild. 2012. "Oil, Islam, Women, and Geography: A Comment on Ross (2008)." *Quarterly Journal of Political Science* 7.1: 69–87.

Hamad, Karim. 2006. "بعد إعلان النتائج النهائية: سيطرة دينية على المجلس" ["After the Announcement of the Final Results: Religious Control over the Council"]. *Akhbar Al-Khaleej*, December 4.

HAQ: Movement of Liberties and Democracy—Bahrain. 2007. "Motivated Change of Demography: Infringements of Political Rights and Inadequate Living Standards." Report submitted to the Universal Periodic Review Working Group of the United Nations Office of the High Commissioner for Human Rights. November 19. Available at http://lib.ohchr.org/HRBodies/UPR/Documents/Session1/BH/MLD _BHR_UPR_S1_2008_Movement ofLibertiesandDemocracyHAQ_%20uprsub mission.pdf.

Heckman, James J. 1976. "The Common Structure of Statistical Models of Truncation, Sample Selection, and Limited Dependent Variables and a Simple Estimator for Such Models." *Annals of Economic and Social Measurement* 5(4): 475–492.

Holes, Clive. 2005. "Dialect and National Identity: The Cultural Politics of Self-Representation in Bahraini *Musalsalāt*." In Paul Dresch and James Piscatori, eds., *Monarchies and Nations: Globalization and Identity in the Arab States of the Gulf*. London: I. B. Tauris.

Horowitz, Donald L. 1985. *Ethnic Groups in Conflict*. Berkeley: University of California Press.

Human Rights Watch. 2004. "Bahrain: Activist Jailed after Criticizing Prime Minister." September 28. Available at www.hrw.org/english/docs/2004/09/29/bahrai9413.htm.

Husain, 'Abd al-Wahhab. 2009. Personal interview. Bahrain. May 31.

Husain, Jasim. 2009. Personal interview. Bahrain. April 16.

International Crisis Group. 2005. "Bahrain's Sectarian Challenge." *Middle East Report No. 40.* May 6.

———. 2011. "Popular Protest in North Africa and the Middle East (VIII): Bahrain's Rocky Road to Reform." *Middle East Report No. 111.* July 28.

Jaggers, Keith, and Ted Robert Gurr. 1995. "Tracking Democracy's Third Wave with the Polity III Data." *Journal of Peace Research* 32.4: 469–482.

Jamri, Mansour al-. 1998. "State and Civil Society in Bahrain." Paper presented at the Annual Conference of the Middle East Studies Association. Chicago. December 9.

Katzman, Kenneth. 2013. "Bahrain: Reform, Security, and U.S. Policy." Washington, D.C.: U.S. Congressional Research Service. April 1.

———. 2013. "Kuwait: Security, Reform, and U.S. Policy." Washington, D.C.: U.S. Congressional Research Service. March 29.

Khalaf, Abdulhadi. 2000. "The New Amir of Bahrain: Marching Sideways." *Civil Society* 9.100: 6–13.

———. 2007. "Al Khalifa, Hamad bin Isa (1950–)." In Michael R. Fischbach, ed., *The Biographical Encyclopedia of the Modern Middle East and North Africa.* Farmington Hills, Mich.: Thomson Gale.

———. 2008. "The Outcome of a Ten-Year Process of Political Reform in Bahrain." *Arab Reform Brief No. 24.* Available at www.arab-reform.net/sites/default/files/ARB.23 _Abdulhadi_Khalaf_ENG.pdf

Khawajah, 'Abd al-Hadi al-. 2009. "تضحيات الحسين تفضح 'العصابة الحاكمة' وتسقطها من الحكم" ["How the Sufferings of al-Husain Exposed 'the Ruling Gang' and Toppled It from Power"]. Unpublished address delivered in Manama near the al-Khawajah Mosque. January 7.

———. 2009. Personal interview. Bahrain. April 29.

Khomeini, Ruhollah. 2002 [1963]. *Islam and Revolution.* Trans. Hamid Algar. New York: Kegan Paul.

Khuri, Fuad. 1980. *Tribe and State in Bahrain.* Chicago: University of Chicago Press.

Kingdom of Bahrain. 2010. *Annual Economic Review.* Manama, Bahrain: Economic Development Board.

———. 2010. "Population of the Kingdom of Bahrain by Governorate, Nationality & Sex—2010." Isa Town, Bahrain: Central Informatics Organization.

———. 2011. *Census and Demographic Statistics.* Isa Town, Bahrain: Central Informatics Organization.

———. 2014. "Royal decree demarcates electoral districts, constituencies and electoral subcommittees." Information Affairs Authority. September 23, 2014. Available at http://www.iaa.bh/news-details.aspx?id=463.

Kinninmont, Jane. 2011. "Framing the Family Law: A Case Study of Bahrain's Identity Politics." *Journal of Arabian Studies* 1.1: 53–68.

———. 2012. "Bahrain: Beyond the Impasse." London: Royal Institute for International Affairs.

Law, Bill. 2007. "Riots Reinforce Bahrain Rulers' Fears." *Sunday Telegraph,* July 22.

Lawson, Fred H. 2004. "Repertoires of Contention in Contemporary Bahrain." In Quintan Wiktorowicz, ed., *Islamic Activism: A Social Movement Theory Approach.* Bloomington: Indiana University Press.

Louër, Laurence. 2008. "The Political Impact of Labor Migration in Bahrain." *City & Society* 20.1: 32–53.

———. 2008. *Transnational Shia Politics*. New York: Columbia University Press.

Luciani, Giacomo. 1987. "Allocation vs. Production States: A Theoretical Framework." In Hazem Beblawi and Giacomo Luciani, eds., *The Rentier State: Nation, State and Integration in the Arab World*. Vol. 2. London: Croom Helm.

Mahdavy, Hossein. 1970. "Patterns and Problems of Economic Development in Rentier States: The Case of Iran." In M. A. Cook, ed., *Studies in the Economic History of the Middle East: From the Rise of Islam to the Present Day*. London: Oxford University Press.

Marzuk, Khalil Ibrahim al-. 2009. Personal interview. Bahrain. April 30.

Matthiesen, Toby. 2012. "Saudi Arabia's Shiite Escalation." *Foreign Policy*, July 10.

———. 2013. *Sectarian Gulf: Bahrain, Saudi Arabia, and the Arab Spring That Wasn't*. Palo Alto, Calif.: Stanford University Press.

Michalski, Bernadette. 1996. *The Mineral Industry of Bahrain*. Washington, D.C.: United States Geological Survey.

Mitchell, Jocelyn. 2013. "Beyond Allocation: The Politics of Legitimacy in Qatar." PhD diss., Georgetown University.

Okruhlik, Gwenn. 1999. "Rentier Wealth, Unruly Law, and the Rise of Opposition: The Political Economy of Oil States." *Comparative Politics* 31.3: 295–315.

Peterson, J. E. 2002. "Bahrain's First Reforms under Amir Hamad." *Asian Affairs* 33.2: 216–227.

———. 2004. "Bahrain: The 1994–1999 Uprising." *Arabian Peninsula Background Note, No. APBN-002*. Available at www.JEPeterson.net.

———. 2008. "The Promise and Reality of Bahraini Reforms." In Joshua Teitelbaum, ed., *Political Liberalization in the Gulf*. New York: Columbia University Press.

Qambar, Samy 'Ali Hasan. 2009. Personal interview. Bahrain. May 17.

Qubain, Fahim I. 1955. "Social Classes and Tensions in Bahrain." *Middle East Journal* 9.3: 269–280.

Ross, Michael L. 2001. "Does Oil Hinder Democracy?" *World Politics* 53.3: 325–361.

———. 2008. "Oil, Islam, and Women." *American Political Science Review* 102.1: 107–123.

———. 2009. "Oil and Democracy Revisited." Unpublished manuscript. March 2. Available at www.sscnet.ucla.edu/polisci/faculty/ross/Oil and Democracy Revisited.pdf.

Rumaihi, Muhammad Ghanim al-. 1978. *Bahrain: Social and Political Change since the First World War*. London: Bowker.

Sa'idi, Jasim Ahmad al-. 2009. Personal interview. Bahrain. May 14.

Salamé, Ghassan. 1990. "'Strong' and 'Weak' States: A Qualified Return to the *Muqaddimah*." In Giacomo Luciani, ed., *The Arab State*. London: Routledge.

Saleh, Ali Abdullah. 2010. "في حديثه لـ 'واجه الصحافة'مع داود الشريان، الرئيس اليمني: لا وجود أمريكياً في أراضينا ولن أترشح للرئاسة" ["In His Interview with 'Meet the Press' with the Yemeni president Dawud al-Sharyan: 'There is not a Single American on Our Soil, and I Won't Run for the Presidency.'" Interview broadcast on Al-Arabiya. March 19. Available at www.alarabiya.net/articles/2010/03/19/103454.html.

Sambidge, Andy. 2012. "Dubai Ruler Announces New Mega City Project." *Arabian Business*, November 24.

Sharif, Ebrahim. 2009. Personal interview. Bahrain. May 11.

Shehabi, Ala'a. "Why is Bahrain Outsourcing Extremism?" *Foreign Policy*. October 29, 2014.

Sick, Gary G. 1997. "The Coming Crisis in the Persian Gulf." In Gary G. Sick and Lawrence G. Potter, eds., *The Persian Gulf at the Millennium: Essays on Politics, Economy, Security, and Religion*. New York: St. Martin's Press.

Sigelman, Lee, and Langche Zeng. 1999. "Analyzing Censored and Sample-Selected Data with Tobit and Heckit Models." *Political Analysis* 8.2: 167–182.

Singace, 'Abd al-Jalil al-. Personal interview. Bahrain. April 2009.

Skocpol, Theda. 1982. "Rentier State and Shi'a Islam in the Iranian Revolution." *Theory and Society* 11.3: 265–283.

Slackman, Michael. 2009. "In a Landscape of Tension, Bahrain Embraces Its Jews. All 36 of Them." *New York Times*, April 5.

———. 2009. "Sectarian Tension Takes Volatile Form in Bahrain." *New York Times*, March 28.

Spindle, Bill, and Margaret Coker. 2011. "The New Cold War." *Wall Street Journal*, April 16.

Toumi, Habib. 2006. "Bahrain's Islamist MP Calls for Removal of Sectarian Banners." *Gulf News*, February 19.

———. 2006. "Religious Fatwas Used to Explain Poll Participation." *Gulf News*, November 21.

———. 2007. "Minister: Don't Use Religious Events to Fuel Sectarianism." *Gulf News*, February 15.

———. 2009. "Call to Lift Bahrain Parliamentary Immunity over Offensive Remarks." *Gulf News*, February 26.

Trabelsi, Habib. 2002. "Bahrain's Shiite Muslims Cry Foul over Dual Nationality Plan." *Khaleej Times*, June 16.

Ulrichsen, Kristian Coates. 2012. "The UAE: Holding Back the Tide." *Open Democracy*, August 5.

———. 2013. "Academic Freedom and UAE Funding." *Foreign Policy*, February 25.

U.S. Central Intelligence Agency. "Bahrain—Population Density." 1993. *Atlas of the Middle East*. Washington, D.C.: U.S. Central Intelligence Agency. Available at www.lib.utexas.edu/maps/atlas_middle_east/bahrain_pop.jpg.

Valeri, Marc. 2011. "The Qaboos-State under the Test of the 'Omani Spring': Are the Regime's Answers Up to Expectations?" *Les dossiers du CERI*. Paris: Sciences Po.

———. 2014. "Identity Politics and Nation-Building under Sultan Qaboos." In Larry Potter, ed., *Sectarian Politics in the Persian Gulf*. London/New York: Hurst/Oxford University Press.

Vance, Colin. 2006. "Marginal Effects and Significance Testing with Heckman's Sample Selection Model: A Methodological Note." *RWI Discussion Papers No. 39*. RWI Essen. Available at http://repec.rwi-essen.de/files/DP_06_039.pdf.

Vandewalle, Dirk. 1987. "Political Aspects of State Building in Rentier Economies: Algeria and Libya Compared." In Hazem Beblawi and Giacomo Luciani, eds., *The Rentier State: Nation, State and Integration in the Arab World*. Vol. 2. London: Croom Helm.

Wehrey, Frederic. 2013. *Sectarian Politics in the Gulf: From the Iraq War to the Arab Uprisings*. New York: Columbia University Press.

Williams, Melissa S. 2010. *Voice, Trust, and Memory: Marginalized Groups and the Failings of Liberal Representation*. Princeton, N.J.: Princeton University Press.

World Trade Press. "Population Density Map of Bahrain." 2007. *Best Country Reports*. Petalumas, CA: World Trade Press. Available at www.bestcountryreports.com/Population_Map_Bahrain.html.

Wright, Robin, and Peter Baker. 2004. "Iraq, Jordan See Threat to Elections from Iran." *Washington Post*, December 8.

Wright, Steven M. 2008. "Fixing the Kingdom: Political Evolution and Socio-economic Challenges in Bahrain." Doha: Center for International and Regional Studies.

Worrall, James. 2012. "Oman: The 'Forgotten' Corner of the Arab Spring." *Middle East Policy* 19.3: 98–115.

Worth, Robert, and Nada Bakri. 2008. "Hezballah Ignites a Sectarian Fuse in Lebanon." *New York Times*, May 18.

Yates, Douglas A. 1996. *The Rentier State in Africa: Oil Rent Dependency and Neocolonialism in the Republic of Gabon*. Trenton, N.J.: Africa World Press.

Zaher, Mohammed 2009. "GCC: Fiscal Stimulus and Reforms Are Optimal Choice under Current Circumstances." *GCC Research Note*. National Bank of Kuwait (NBK). April 2. Available at www.kuwait.nbk.com.

Index

Note: page numbers in italics refer to figures or tables.

About the Author

JUSTIN GENGLER is Senior Researcher at the Social and Economic Survey Research Institute, Qatar University. He received his PhD in Political Science from the University of Michigan in 2011. He is a contributor most recently to *The Arab Revolts* (Bloomington: Indiana University Press, 2013) and *Sectarian Politics in the Persian Gulf* (London/New York: Hurst/Oxford University Press, 2014). His work has appeared in *Middle East Policy*, *Journal of Arabian Studies*, *Foreign Affairs*, *Foreign Policy*, and the Middle East Research and Information Project.

www.ingramcontent.com/pod-product-compliance
Lightning Source LLC
Chambersburg PA
CBHW070325270326
41926CB00017B/3759

* 9 7 8 0 2 5 3 0 1 6 8 0 5 *